M. Townsend

MW00801066

Introduction to Fine Ceramics

Introduction to Fine Ceramics

Applications in Engineering

Edited by
Noboru Ichinose
Waseda University, Japan

Authors: Noboru Ichinose
Katsutoshi Komeya
Naohiko Ogino
Akihiko Tsuge
Yuuji Yokomizo

Translated by Keizo Hisatake
and Charles G. Aschmann

JOHN WILEY & SONS LTD
Chichester · New York · Brisbane · Toronto · Singapore

This edition © 1987 N. Ichinose

Original Japanese Edition

Introduction to Fine Ceramics by Noboru Ichinose
First published in 1983 by Ohmsha, Ltd.
Copyright © 1983 by N. Ichinose
English translation rights arranged with Ohmsha, Ltd.

All rights reserved.

No part of this book may be reproduced by any means, or transmitted, or translated into a machine language, without the written permission of the publisher.

Library of Congress Cataloging in Publication Data:

Zukai fain sermaikusu tokuhon. English.
  Introduction to fine ceramics.

  Translation of: Zukai fain sermikusu tokuhon.
  Includes index.
  1. Ceramics. I. Ichinose, Noboru. II. Title.
TP807.Z8513 1987   666   86-32484

ISBN 0 471 91445 2 (U.S.)

British Library Cataloguing in Publication Data:

Introduction to fine ceramics.
  1. Ceramics
  I. Ichinose, Noboru II. Hisatake, Kiezo
  III. Aschmann, Charles G.
  666   TP807

ISBN 0 471 91445 2

Typeset by KEYTEC, Bridport, Dorset
Printed in Great Britain by Anchor Brendon Ltd.

Preface

Recently, 'new ceramics' or 'fine ceramics' has become a topic of discussion. Fine ceramics, especially, has been in the newspapers and weeklies, since the Ministry of International Trade and Industry decided in 1981 to promote and develop it as a part of its 'Next Generation Industries Technology Development Program,' with investments of around ¥ 14 billion (US$ 58.3 million) over ten years. In addition, makers and users of fine ceramics formed the Fine Ceramics Society in July 1981. The goals of this society are promotion of research and information exchange in manufacture and utilization, market research, technology development, positive international exchange, and improvements in the fundamentals of the fine ceramics industry. Thus the term 'fine ceramics' has become better-known, and has promoted a discussion about its exact meaning. Currently there is no exact definition and, according to the user, 'fine' could mean anything from little, delicate or beautiful to high level.

Tentatively, then, I will give a narrow definition. Ceramics have many properties, but one can think of ceramics that have special practical properties, as well as those with specific unusual properties, as fine ceramics. One can say that fine ceramics is functional ceramics.

Accordingly, one can see that, even though there have been many books specializing in fine ceramics lately, they are quite difficult. There have been almost no illustrative, easy-to-understand books on the subject. This book uses a question and answer format to provide just that kind of explanation of fine ceramics. The book is organized as follows: Chapter 1, Fundamentals of Ceramics: Questions and Answers; Chapter 2, Structural Ceramics: Questions and Answers; Chapter 3, Electronic Ceramics: Questions and Answers; Chapter 4, Glass and Optical Fibers: Questions and Answers; and finally, Chapter 5, New Technology of Ceramics: Questions and Answers. In this way the book tries to give the reader a concrete picture, including tables and diagrams, of fine ceramics from the basics to the most recent research.

In the future it seems that fine ceramics will continue to develop. And if this book helps engineers and materials researchers to a deeper understanding or helps pave the way for new engineering or materials development even a little, the writers will be gratified.

Finally, we wish to express our appreciation and thanks to the publishers, Ohmsha, Ltd and John Wiley & Sons Ltd. Their cooperation and hard work have been essential to its completion. Also, a personal note, I was kindly allowed to dedicate the book to my beloved daughter who died at the age of thirteen while I was compiling the original manuscript.

Noboru Ichinose

Contents

1 Fundamentals of Ceramics: Questions and Answers

Chapter 1 will touch upon the fundamentals of ceramics. The differences between ceramics, metals and organic materials, microstructures, and grain boundaries will be mentioned. Powder synthesis, forming processes, and sintering processes will be explained. Sintering processes such as hot pressing and HIP will be commented on. Why ceramics are translucent, why thay are strong at high temperatures, and some more recent topics will be touched upon.

1.1 WHAT ARE CERAMICS AND HOW DO THEY DIFFER FROM METALS AND ORGANIC MATERIAL?

In our society, industrial products are classified in three categories: metals like iron, copper, and aluminum; organic materials like epoxy resins and rubber; and ceramics like porcelain, refractories, and electronics wares (Fig. 1.1).

The term 'ceramics' comes from *keramos*, the ancient Greek word for objects made of fired clay. While retaining this original meaning, the word has also come to designate one of the three main material categories. It has taken on much of what was once included in 'brickmaking', 'pottery', and 'glassmaking'.

What are the main features of the three kinds of materials, and what are the differences between ceramics and the others? Basically, they are a product of the differences in chemical bonding. The consequences are remarkably different physical and chemical properties, as well as different manufacturing processes. Let us compare these materials in more detail.

1

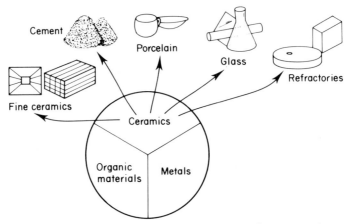

Fig. 1.1 Ceramics: one of the three main product categories

(1) Metals

Chemical bonding in metals, or metallic bonding, is a binding of atoms to other atoms by free electrons. There is good heat and electrical conductivity, non-penetration and reflection of light, and toughness, all of which can be explained by the existence of these mobile electrons.

(2) Organic Materials

Organic materials are compounds of carbon, hydrogen, and oxygen which are found in living organisms, regardless of molecule size. A chain of carbon atoms is formed, to which hydrogen and oxygen are added. The chemical bonding between these non-metallic elements, through which groups of molecules are formed, is usually called covalent bonding. The force, van der Waals force, holding the molecules together is weak. Consequently, the melting points are low, and these materials are characteristically easily shaped and processed.

(3) Ceramics

Almost all ceramics are compounds of the electropositive and electronegative elements of the periodic table. Mostly the bonding is ionic, but in a few cases, covalent or metallic bonding occurs. There are also numerous configurations for each combination of elements, and therefore, various material functions. The common features of these materials are:

(a) high heat resistance
(b) electrically insulating or semiconducting with various magnetic and dielectric properties

(c) strong resistance to deformation, brittle fracture
(d) low toughness.

These properties are sometimes advantageous and sometimes not. Yet technical developments have tended to make use of the good properties and overcome the bad ones. New ceramics have been developed, studied, and kept in the scientific limelight. These are what have been called 'new ceramics' or 'fine ceramics'.

In Table 1.1, ceramics are classified according to application. Although, in the strictest sense, glass should be separated from ceramics, it is included here in concurrence with the above definition. A further division could also be made, between the ceramics that have been used for centuries (traditional ceramics) and the new attention-getting functional materials and machine materials (fine ceramics).

Table 1.1 Ceramics groups (following the *Ceramic Engineering Handbook*)

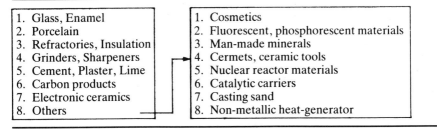

1. Glass, Enamel	1. Cosmetics
2. Porcelain	2. Fluorescent, phosphorescent materials
3. Refractories, Insulation	3. Man-made minerals
4. Grinders, Sharpeners	4. Cermets, ceramic tools
5. Cement, Plaster, Lime	5. Nuclear reactor materials
6. Carbon products	6. Catalytic carriers
7. Electronic ceramics	7. Casting sand
8. Others	8. Non-metallic heat-generator

1.2 WHAT IS FINE CERAMICS?

Usually, ceramics are oxides or non-oxides composed of metallic and non-metallic elements (excluding carbon). Therefore, there are many kinds of ceramics with functions in many fields. In Fig. 1.2 and Table 1.2, the concepts of fine ceramics and the function–material–applied product relationships of oxides and non-oxides are shown. A study of these charts should provide an adequate summary of the many functions of ceramics.

Recently it has been said that 'fine ceramics' is used to designate the ceramics which have high additive value. What, then, is fine ceramics? The specialized functions exhibited by the materials in Table 1.2 cannot be obtained by simply pressing and sintering unrefined raw material. It is necessary to synthesize ceramics using highly refined raw material, rigorously controlled composition, and strictly regulated forming and sintering. The ceramics obtained by this type of process are called 'fine ceramics'. However, this term is not yet completely fixed in meaning. In a broad sense, all of the materials in Table 1.2 are fine ceramics, but in a narrow sense, fine ceramics is limited to the ceramics used as machine materials. Even now disputes as to what to include continue.

Table 1.2 Functions and practical applications of ceramics

Functional group	Oxide ceramics			Non-oxide ceramics		
	Function	Materials	Application	Function	Materials	Application
Electric, electronic functions	Insulating	Al_2O_3, BeO	Substrates	Insulation	C, SiC	Substrates
	Dielectrical	$BaTiO_3$	Capacitor	Electrical conductivity	SiC, $MoSi_2$	Heat generator
	Piezo-electronics	$Pb(Zr_2, Ti_{1-2})O_3$, ZnO, SiO_2	Oscillator, ignition junction, Surface elastic wave delaying junction	Semi-conductivity	SiC	Varistor, lightning shunt
	Magnetism	$Zn_{1-2}Mn_2$	Memory, operation junction	Electron emission	LaB_6	Electron gun thermal anodes
		Fe_3O_4	Magnetic core			
	Semi-conductivity	SnO_2,	Gas-sensor			
		$ZnO-Bi_2O_3$ $BaTiO_3$	Varistor Resistance junction			
	Ionic conductivity	β-Al_2O_3,	NaS battery			
		Stable ZrO_2	Oxide-sensor			

Table 1.2 (*contd.*)

Functional group	Oxide ceramics			Non-oxide ceramics		
	Function	Materials	Application	Function	Materials	Application
Mechanical functions	Wear resistance	Al_2O_3, ZrO_2	Polishing materials Grindstones	Wear resistance	B_4C, diamond	Wear-resistant materials, grindstones
	Machineability		Cutting tools	Machineability	C–BN, TiC, WC, TiN	Cutting tools
				Strength functions	Si_3N_4, SiC	Engines, heat resistors
				Lubricating functions	Sialon C, MoS_2, h-BN	Anticorrosives, tools Solid lubricants, mold-releasing agents
Optical functions	Fluorescence	Y_2O_2S: Eu	Fluorescent materials	Transparency	AlON, nitrogen glass	Windows
	Transparency	Al_2O_3	Sodium lamp Mantle tube	Reflectiveness	TiN	Light collectors
	Optical polarization	PLZT	Optical polarization junction			
	Light conductivity	SiO_2, multiple-constituent type glass	Optical communication fibers			

Table 1.2 (*contd.*)

Functional group	Oxide ceramics			Non-oxide ceramics		
	Function	Materials	Application	Function	Materials	Application
	Heat resistance	Al_2O_3	Structural refractories	Heat resistance	SiC, Si_3N_4, h-BN, C	Various refractories
Thermal functions	Heat insulation	$K_2O \cdot nTiO_2$, $CaO \cdot nSiO_2$, ZrO_2	Heat-insulating materials	Heat insulation	C, SiC	Various heat insulators
	Heat conductivity	BeO	Substrates	Heat conductivity	C, SiC	Substrates
Nuclear power related functions	Nuclear reactor materials	UO_2 BeO	Nuclear fuel Moderator	Nuclear reactor materials	UC C, SiC C B_4C	Nuclear fuel Coated nuclear fuel Moderator Control rod material
Biochemical functions	Teeth & bone materials	Al_2O_3, $Ca_5(F,Cl)P_3O_{12}$	Artificial teeth & bones	Corrosion resistance	h-BN, TiB_2 Si_3N_4, Sialon	Evaporation chamber Pump materials, others
	Carrier ability	SiO_2, Al_2O_3	Catalytic carriers		C, SiC	Anticorrosives

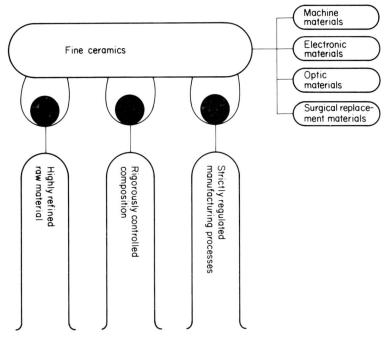

Fig. 1.2 General concepts of fine ceramics

The above definition is limited to sintered ceramics, but in Table 1.2, single crystals, thin film, and glass are also included. It is somewhat contradictory to include these, but they are included because of their functional applications. Also, besides, 'fine ceramics', many other terms have been coined to describe these materials: 'new ceramics,' 'special ceramics,' 'modern ceramics,' 'engineering ceramics,' 'electroceramics,' etc., have all been used. Lack of standardization has caused much complexity, but fine ceramics technology is that which gives ceramics additive value. And in their manufacture it is important to control the powder and the microstructure. The rest of this book deals with a number of techniques necessary for the production of high-quality fine ceramics.

1.3 WHAT IS THE CRYSTAL STRUCTURE OF CERAMICS

Ceramics have inorganic structures. Sometimes there are glass-like amorphous structures, but almost all ceramics have a crystal structure. Here, then, is an outline of structures.

Matter can exist in three different states: solid, liquid, or gas (Fig. 1.3). These states are distinguished by interatomic distances. The atoms or molecules in a gas are diffused in space. As they become liquid and then solid, the regularity of their arrangement is increased. In the solid state, the attractive force overcomes the thermal effect which separates the atoms, and they begin to occupy fixed sites.

8

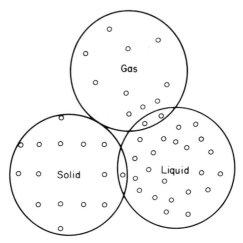

Fig. 1.3 Molecular positions in the three states of matter

In order to understand crystal structure, it is necessary to know something about atomic structure, but explaining atomic structure in detail is not within the scope of this section, so it will only be touched upon in relation to bonding strength in ceramics. There are ionic bonds, covalent bonds, van der Waals bonds, and metallic bonds. Oxide ceramics mainly have ionic bonding. Covalent bonding, which is strongly directional, can be seen in non-oxides, silicon, etc. The crystal structure of oxides is governed by the close-packing of O^{2-} ions, with cations situated among them. In this way polyhedra are formed with cations surrounded by O^{2-} ions. The number of anions surrounding a cation is called the 'coordination number.' This number is determined geometrically by the ratio of the ionic radii. As a result, calculated predictions closely agree with the real polyhedra. Triangular, tetrahedral, rhombohedral, octahedral, cubic, and close-packed configurations of anions exist. The real structure is formed spreading out in three dimensions.

Some minerals which are identical in composition have different crystal structures and different chemical and physical properties. Such minerals are said to be 'dimorphic.' The assumption of two or more crystal structures by the same substance is called 'polymorphism.' For example, there are: quartz, cristobalite, and tridymite configurations of SiO_2; rutile, anatase, etc., configurations of TiO_2; and diamond and graphite configurations of carbon (C). The polymorphic crystal structures are called 'modifications' of a substance, and the change from one crystal structure to another is known as 'transformation.' It is helpful to learn these terms because they crop up all the time.

Several typical examples of oxide structure groups are: rock salt (NaCl), cubic; zincblende (β-ZnS), cubic; wurtzite (α-ZnS), hexagonal; fluorite (CaFe), cubic; rutile (TiO_2), tetragonal; corundum (α-Al_2O_3), hexagonal; perovskite ($CaTiO_3$); spinels like magnesium aluminate spinel ($MgAl_2O_4$)

(see Fig. 1.4). Each structural group has similar chemical and physical properties, and these groups are the materials known as ceramics.

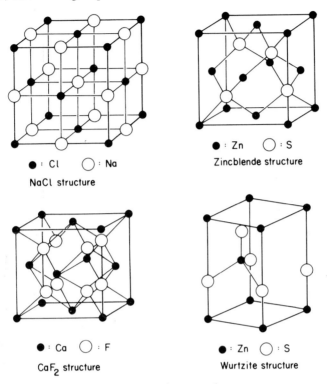

● : Cl ○ : Na

NaCl structure

● : Zn ○ : S

Zincblende structure

● : Ca ○ : F

CaF$_2$ structure

● : Zn ○ : S

Wurtzite structure

Fig. 1.4 Representative crystal structures

1.4 WHAT IS THE MICROSTRUCTURE OF CERAMICS

The microstructure of polycrystalline ceramics is usually complex, as is shown by Fig. 1.5, and distinguished by the existence of grain boundaries, which are not seen in single crystals. Also, the existence of pores, imperfections, and multiphase compositions makes for great variety. Up to now, grain boundaries and additional phases were thought to be undesirable, and the goal was to eliminate them and obtain a structure as close to single crystals as possible. However, new processes have been found that make positive use of these surfaces and grain boundaries, and functional ceramics in which these properties are important are developing rapidly.

Fig. 1.5 can be used to explain what is meant in terms of grain boundaries. In the grain boundary region, energy is increased, so impurities tend to gather there. The impurities exist as a second or third phase among the constituent particles or segregate into the grain boundaries. With an increase in the amount of impurities and additives, the microstructure shifts from (a) to (c).

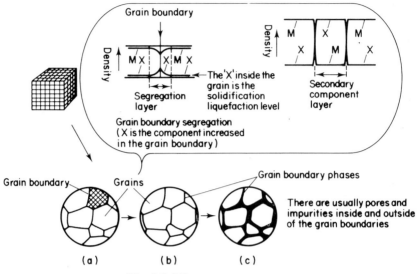

Fig. 1.5 Microstructure pattern

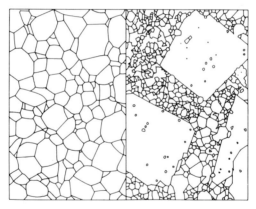

Fig. 1.6 Microstructure of alumina. (a) Normal grain growth. (b) Exaggerated grain growth

In such a case, the shapes of the grain boundaries or crystallites depend on the material, its constituents, and the sintering process. For example, when alumina is produced by sintering an Al_2O_3—MgO system, MgO segregates in the grain boundary, and suppresses grain growth. The pores in the sintered structure, as shown in Fig. 1.6(a) can be eliminated. This is a transparent alumina which is used in the envelope tubes of sodium lamps.

Fig. 1.6(b) shows exaggerated grain growth in alumina. It is said to depend on the existence of impurities and the size of particles in the raw material.

From the point of view of strength, grain-size is an important factor. In general the strength of a material (σ_f) is defined by the following equation:

$$\sigma_f = \frac{1}{Y} \sqrt{\frac{2v_i E}{C}}$$

Here, v_i equals the fracture energy, E is the Young modulus, C equals half the major axis of the crack, and Y is a geometrical constant. C corresponds to the largest crack, which is directly related to strength. Its maximum size is related to the grain-size and the pore-size; therefore, the results of both exaggerated grain growth and large average grain-size are a lowering of strength. Thus, as in the alumina mentioned above, to guarantee strength, it is necessary to minimize the grain-size. The above equation corresponds to the fracture toughness value, K_{1c}. In ceramic silicon nitrides, which are noted as a machine material, the shape of the crystallites is elongated to raise K_{1c} (Fig. 1.7).

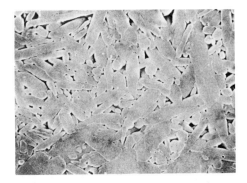

Fig. 1.7 Si_3N_4-Y_2O_3 system microstructure after sintering

Also, in electronic ceramics, the existence of grain boundaries is very important. They are utilized in zinc oxide (ZnO) varistors, PTC ceramics, BL condensers, etc. On the other hand, high-porosity ceramics, in which the surfaces are used, has wide applications in products like catalytic carriers, gas-sensors, and moisture-sensors.

1.5 WHAT ARE GRAIN BOUNDARIES

In general, ceramics are produced by shaping powdered raw material and sintering it. Ceramics obtained in this way are polycrystalline, an aggregation of fine crystalline grains, and grain boundaries inevitably exist. They play an important role in the sintering process, and have a large influence on chemical and physical properties. They were once thought to cause bad effects in polycrystalline materials, but more recently grain boundaries have been the characteristic most commonly used in electronic devices. Their use has revolutionized the field. The 'can also be done with ceramics' era has given way to the 'can only be done with ceramics' era.

As shown in Fig. 1.8(a), the microstructure of ceramics contains: fine crystalline grains, grain boundaries, impurities segregated in the grain boundaries, pores in the grain boundaries, impurities within the grains, and pores in the grains. The grains, or fine particles which are the main constituent of ceramics, range from one micrometer to tens of micrometers in size, and the directions of their axes are arbitrary. The size of the grains depends on the size of the particles in the raw material, on impurities, and on the sintering conditions.

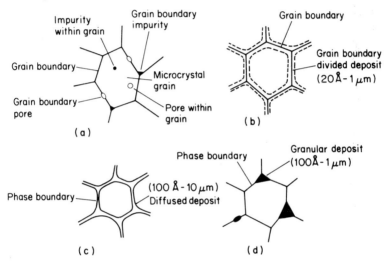

Fig. 1.8 (a) Representative ceramic microstructure. (b) Grain boundary divided deposit. (c) Diffused deposit. (d) Granular deposit

Within ceramic grain boundaries, there are crystal lattice defects such as dislocations and pores, as well as crystal lattice deformations. Correspondingly, as is shown in Fig. 1.8(b), (c), and (d), impurities tend to gather in these areas and form grain boundary divided deposits, diffused deposits, and granular deposits.

(1) Grain boundary divided deposits

In general, the ionic stratifications of impurities which are separated along the grain boundary are called 'grain boundary divided deposits.' They range in thickness from 20 Å to 1 μm. In the grain boundary, impurities are very easily dissolved, and therefore, their crystal phases are considered to be very different from those inside the grains.

(2) Diffused deposits

When the amount of impurities is large, higher than the saturation point of the solution, they are precipitated into the grain boundary in a separate

crystalline phase. These precipitates are either diffused or granular. Generally, diffused deposits are brought on by liquid phase sintering.

   Liquid phase sintering occurs when the melting point of the precipitates in
the grain boundaries is lower than the sintering temperature of the ceramics.
If there is good or complete wetting, the liquid flows completely into the grain
boundary and surrounds each of the fine particles, forming a diffused deposit.
The ZnO varistor is a typical example.

(3) Granular deposits

When the amount of impurities is much greater than the saturation point of
the solution, and the melting point is higher than the sintering temperature,
particle-like impurities may be precipitated in the grain boundary. In
transparent alumina ceramics, MgO is used as an additive, and when the
amount added is large, $MgAl_2O_4$ is precipitated in the grain boundaries,
making transparency lower.

   The physical and chemical phenomena peculiar to grain boundaries are:
(a) grain boundary diffusion
(b) control of the formative reaction mechanism with respect to the grain
    boundary
(c) grain boundary potential
(d) high resistance in the grain boundary
(e) grain boundary bonding.

Fig. 1.9 illustrates these phenomena.

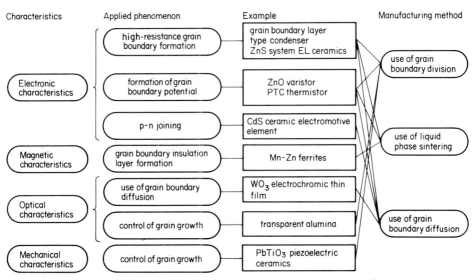

Fig. 1.9 Examples of applications of grain boundary phenomena

1.6 STANDARD CERAMICS ARE POLYCRYSTALLINE; HOW ARE THEY MADE?

Since polycrystalline ceramics are made by sintering, they are usually referred to as sintered ceramics. But there are many kinds of ceramics, because there is a wide variety of compositions and forming processes. One example of the many forming processes is illustrated inside the front cover of this book. Raw material is selected with respect to the desired product. If necessary, it is ground to fine particles, or it is mixed in its original form. It is fired in the desired form, and follow-up manufacturing processes are completed. A look at alumina substrates used in integrated circuits will allow a more thorough explanation of ceramic manufacturing processes.

High-quality alumina substrates are commonly used in IC (integrated circuit) laminations, because of the necessity for protection from mechanical and chemical erosion caused by exposure to outside air and the requirements for heat radiation. The goal is to make an inexpensive plate-like ceramics that is free from electronic leakage. This kind of challenge is an inspiration to industry, irrespective of product. A highly useful product is made, and a unique process is devised.

The manufacturing of alumina substrates begins with the mixing of powdered alumina and additives by weight. A sheet of the powder mixed with organic resin, called a 'green sheet,' has been prepared in advance by screen-printing a conductive line network on its surface, using a paste composed mostly of tungsten powder.

Several of these sheets are then laminated and sintered at over 1,500°C in a hydrogen atmosphere. Prior to sintering, it is necessary to get rid of the organic resin used for forming by heating the system at a low temperature. The formed product, which is mainly composed of alumina powder, is sintered at a high temperature and becomes dense through contraction. Fig. 1.10 shows the change in structure associated with the manufacture of the sintered product. It is an aggregation of particles ranging in size from several micrometers to several tens of micrometers, connected by grain boundaries.

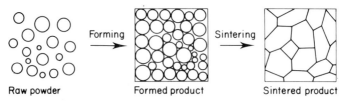

Fig. 1.10 Structural changes accompanying the manufacture of a sintered product

Tungsten is widely used as a conductor since it can resist the high temperatures of sintering; its sintering characteristics are similar to those of alumina, and it has great bonding strength while retaining high conductivity. Therefore, elaborate integrated circuit substrates can be formed using

15

tungsten conductor networks. In such cases, the outer surface of the tungsten is nickel (Ni) plated, and finally, a fine gold plating is added after the leads have been silver waxed.

Alumina has been used as an example, but the manufacturing processes for all sintered ceramics—porcelain, refractories, the new electronic ceramics, and structural ceramics—are the same in principle. When the starting material is a powder, the synthesis, or refining of the powder is one of the most important processes affecting the quality of the final product. Ceramics are different from ordinary metals or plastics in that the powder forming changes diversely with regard to the desired product. This will be commented on in Section 1.8.

Figs. 1.11 and 1.12 show the manufacturing schematics for two typical ceramics.

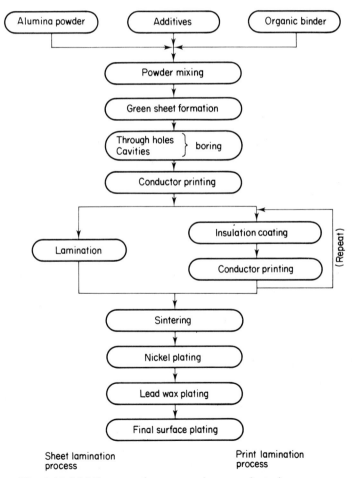

Fig. 1.11 Multilayer package ceramics manufacturing process

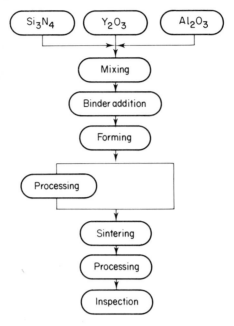

Fig. 1.12 Example of sintered Si_3N_4 production

1.7 HOW IS CERAMIC POWDER SYNTHESIZED?

Up to now porcelain and refractories, etc., have been made using crushed natural minerals, but in order to make use of the specific functions of the new ceramics, high purity and extremely fine powders are necessary before the raw material is sintered. So new methods have been developed for the synthesis of base materials. These methods, which produce refined base powders that comply with the limits summarized in Fig. 1.13. They are divided into the large categories of (i) solid phase synthesis, (ii) liquid phase synthesis, and (iii) gas phase synthesis. Of these, both (ii) and (iii) are important in powder synthesis, and the methods in (i) are often used in reactions of the products of (ii) and (iii). The following is an explanation of the methods found in (ii) and (iii).

(1) Liquid phase synthesis

In the liquid state there are both liquefaction and solution methods. Liquefaction of ceramics is done by plasma jet. The liquid trickle is then atomized, and as a result of solidification, a fine powder is obtained. This method is widely used in the synthesis of metallic powders, but very high temperatures are necessary for the liquefaction of ceramics. At such high temperatures, decomposition and phase separation are apt to occur; accordingly, the method is rarely used. However, synthesis from solution is

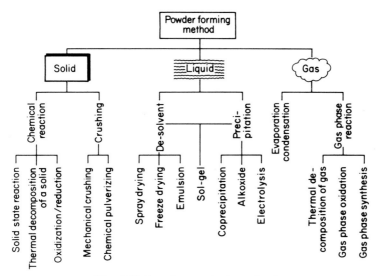

Fig. 1.13 Powder synthesis methods

common. There are two solution methods the de-solvent method and the precipitation method.

(a) De-solvent method

Solvent is physically removed from the solution to increase concentration and precipitate out the solute. In order to concentrate the solution, heating and spray-drying, freeze-drying in a vacuum, or liquid drying, in which the solvent is absorbed by a hygroscopic liquid, may be used. If the solvent has many constituents, a device to prevent their separation is necessary.

(b) Precipitation method

The solute is separated from the solution by precipitation. Included here are co-precipitation, homogeneous precipitation, alkoxide hydrolysis, and electrolysis. The precipitate may be powdered by filtration, washing, drying or thermal decomposition. There is also a method called the sol-gel method. A colloidal solution (sol) with its suspension of dispersed particles may be powdered by allowing it to become a gel. Next, the solvent may either be removed or the solute precipitated out. Thus, this method may be said to belong to both (a) and (b).

(2) Gas phase synthesis

Synthesis from the gaseous state can be done by precipitation of vapor deposition (PVD) or chemical vapor deposition (CVD). Precipitates from gases take on thin film, crystalline particle or fine powder configurations.

(a) PVD

This method is the equivalent of the atomized liquid synthesis method. Raw material is vaporized at high temperatures, and it is condensed to a fine powder by rapid cooling by an arc or plasma.

(b) CVD

This is synthesis through the chemical reaction of the vapor of a metallic compound. Therefore, it is widely used for the formation of SiC and Si_3N_4 (see Fig. 1.14). However, use of the solid synthesis mentioned above has become widespread, and there have been examples of this method of synthesizing Si_3N_4 and SiC recently.

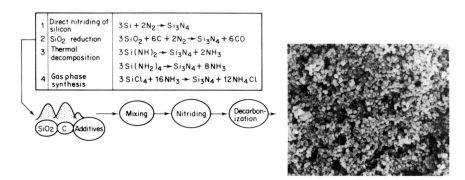

Fig. 1.14 Example of Si_3N_4 powder synthesis

1.8 WHAT ARE THE FORMING PROCESSES OF CERAMICS?

In sintered ceramics, which includes the majority of ceramics, it is difficult to use the melting–pouring–hardening methods employed with metals and plastics, and injection molding requires malleability. There, methods for shaping and then sintering powders are often used. Depending on shape and required characteristics, numerous powder forming processes have been developed. The five main ones are as follows:

(1) die pressing;
(2) rubber mold pressing;
(3) extrusion molding;
(4) slip casting;
(5) injection molding.

However, powder itself consists of solid brittle particles, so it is difficult to fill a die by pressure alone. As the pressure is increased, there is more strain on the compact, and cracks that can cause the compact to fracture are formed. Therefore, a binder is usually added to enhance the fluidity of the powder.

The binders used vary with the different forming processes (1) through (5), and the kind and quantity of binder are determined with respect to the powder used and the product desired. Both hot pressing and HIP (hot isostatic press) are occasionally included in these forming processes, but these are best understood as aids for increasing the external pressure during sintering. Under (4), one can also place film forming, which is widely used in the production of IC substrates (Fig. 1.15).

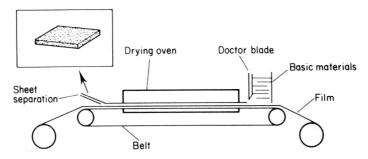

Fig. 1.15 An example of the doctor blade method

(1) Die pressing

Powder is mixed with an organic binder, filled into a die, and pressed into a strong solid product. This method is used for refractories, tile, electronic ceramics, nuclear fuel pellets and other products with relatively simple shapes, and requirements for large numbers. It is an inexpensive method and accurate for close tolerances. Pressures used range from 200 to 2,000 kg/cm^2.

(2) Rubber mold pressing

This method results in a uniform powder compact, and is called rubber mold pressing because of the use of a rubber sheath. Powder is encased in the rubber sheath and is formed in a hydrostatic molding chamber as shown in Fig. 1.16. In this molding chamber, pressure is applied to the powder uniformly, and thus high-quality products are produced. A dry method has also been developed.

(3) Extrusion molding

This method of forming is the extrusion of a stiff plastic mix through a die orifice. This method is largely used for sewer pipe, hollow tile, reactor pipes and other materials having an axis normal to a fixed cross-section. A recent notable example is the ceramic honeycomb used for automobile exhaust purification.

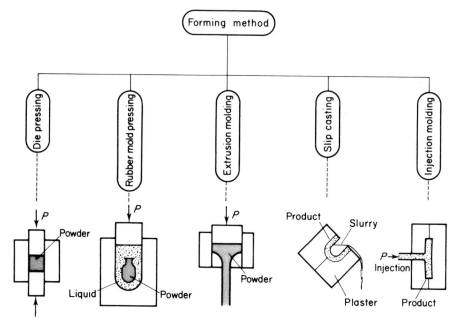

Fig. 1.16 Forming methods

(4) Slip casting

When a suspension tempered with water and other lubricants is poured into a porous plaster mold, water is sucked from the contact area into the mold and a hard layer of clay is built up. This method results in products shaped by the mold's interior surface. Both solid casting and drain casting variations are employed. This method has been used in the manufacture of porcelain since time immemorial.

(5) Injection molding

Plastic is mixed with the powder and the forming process is the same as that for plastics. The advantage of this method is that it allows for the production of complex shapes, but removal of the plastic is difficult because of the large quantity of binder (15–25%) needed. At present, it is not an appropriate method for either large or thick wares.

1.9 WHAT KINDS OF SINTERING PROCESSES ARE THERE?

Ceramics have a wide range of characteristics, so even the polycrystalline structures cannot be obtained through ordinary methods. There are a variety of sintering methods. The main ones are shown in Table 1.3, and a number of concrete examples employing Si_3N_4 are given in Table 1.4. An explanation of each follows.

Table 1.3 Methods for sintering ceramics

● Standard pressure sintering	● Reaction-sintering
● Hot pressing	● Post-reaction sintering
● HIP	● Recrystallization sintering
● Atmospheric pressure sintering	● CVD
● Ultra-high-pressure sintering	

(1) Standard pressure sintering

This category contains what is normally referred to by the word 'sintering.' It is the most widely used. For example, porcelain, refractories, new ceramics such as alumina and ferrites, and many others can be made this way. The advantage of this method is that it is inexpensive compared to other methods because the process is a simple forming of base materials followed by firing. However, in order to make the most of the inherent functions of materials, other methods are preferable occasionally.

(2) Hot pressing

Powder is placed in a mold and sintered under pressure at high temperatures. This method is used with Si_3N_4, SiC, Al_2O_3, etc., but it is expensive; therefore its use is limited.

(3) HIP (see also Section 1.12)

While hot pressing is carried out with pressure along a single axis, this method uses gas pressure to obtain isostatic pressure. This method is used with SiC, Si_3N_4, Al_2O_3, and others, including ultra-high-strength alloys. This is one of the new technologies with great potential for the future.

(4) Ultra-high-pressure sintering

This sintering method is the same as that for the synthesis of diamonds. This sintering process is used for diamond, C-BN, etc., as well as for research materials which are difficult to sinter.

(5) Reaction-sintering

Sintering in which a chemical reaction affects the process is called 'reaction-sintering.' This method is most notably used with Si_3N_4 and SiC. In the case of Si_3N_4, the process can be explained like this: first, a green compact of Si powder is made, and then reaction-sintered in a nitrogen flow. The reaction, $Si + N_2 \rightarrow Si_3N_4$, takes place and Si_3N_4 is obtained as a final sintered

Table 1.4 Various examples of Si₃N₄ sintering processes

Material	Sintering process	Initial material	Sintering accelerator	Manufacturing process	Comments
Si_3N_4	Reaction sintering	Si	none	Si → Forming → Nitrification $\xrightarrow{N}$ → Follow-up processing sintering 1,350–1,600 °C	Sintering shrinkage–0% Extended sintering time Porosity: 13–20%
	Hot pressing	Si_3N_4	MgO Al₂O₃ Y₂O₃ ZrO₂	Si₃N₄ → Forming → Hot pressing → Follow-up accelerator processing 1,700–1,800 °C Pressure: 200–500 kg/cm²	Limited to simple shapes
	Standard pressure sintering	Si_3N_4	MgO Al₂O₃ Y₂O₃	Si₃N₄ → Forming → Firing (sintering) → Follow-up accelerator 1,700–1,800 °C processing in N₂	Sintering shrinkage– 18%
	CVD	SiH₄ or SiCl₄ NH₃	none	SiH₄ or SiCl₄ → Precipitation → Substrate removal NH₃ substrate (C, W, etc.) 800–1,400 °C	Primarily thin products

product. Since N_2 gas is the source of the reaction, high porosity cannot be avoided. The advantage is that there are no dimensional changes.

(6) Post-reaction-sintering

This method was recently developed for Si_3N_4. An example of it is when Y_2O_3 and MgO are mixed in before the nitrization of the Si compact, and after the reaction-sintering, densification is accomplished by means of the additives.

In addition, there is atmospheric pressure sintering, an example of low-pressure HIP, and CVD which is also used as a sintering method.

1.10   WHY IS THE POWDER COMPACT SOLIDIFIED BY SINTERING?

Generally when metallic or ceramic powders are formed and then heated, there is a certain temperature at which they begin to burn, and in most cases there is shrinkage resulting in densification. This process is what is meant by 'sintering.' Thermodynamically, sintering may be thought of as a process for decreasing system-wide energy; that is, it can be understood in terms of kinematic behavior. The driving force for sintering is the surface energy originating in surface stress. The equation,

$$\sigma = \frac{2\gamma}{r}$$

where γ equals the surface tension and r equals the radius of curvature, describes this driving force.

Sintering processes can be divided into three large categories: (i) solid state sintering, (ii) liquid phase sintering, and (iii) gas phase sintering.

As an example, a more in-depth explanation of solid state sintering follows.

The sintering mechanism is usually explained by the equal sphere model shown in Fig. 1.17. Here, the bonding area (neck area) is altered by a transfer of material during the sintering process. In this case, the functional surface stress across the two curved surfaces is expressed by the following equation:

$$\sigma = r \left(\frac{1}{x} - \frac{1}{\rho} \right)$$

Here, since $x \gg \rho$, $\sigma = -\gamma/\rho$ is directed outwards from the neck area. On the other hand, the material transfer may be seen as diffusion. This is when the thermal energy causes a migration of atoms or ions via the lattice vacancies that exist in the material. In a sintering system in which there is a difference in the concentration of lattice vacancies, material goes from the area of lower concentration to that of higher concentration. Depending on the curvature, ρ, in Fig. 1.17, the lattice vacancy density near the neck becomes greater, creating a difference between it and that of the equilibrium area. Consequently, atoms migrate towards the neck area. In addition, since the vapor pressure

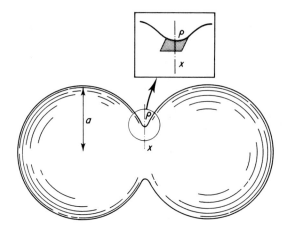

Fig. 1.17 Sintering process sphere model

changes with respect to the curvature, it also gives rise to material transfer.

Material transfer takes place along a number of different paths, and sintering processes are divided accordingly. In the case of solid state sintering, the mechanisms are: (i) volume diffusion, (ii) grain boundary diffusion, (iii) surface diffusion, and (iv) evaporation–condensation.

Theoretically, these are interdependent with the rates of growth or contraction in the neck area. Usually these rates are expressed as a function of time, t, as follows (see Table 1.5):

$$\left(\frac{x}{a}\right)^m \propto t \text{ or } \left(\frac{\Delta L}{L}\right)^n \propto t$$

where x/a is the growth rate and $\Delta L/L$ is the shrinkage rate in the neck area.

Table 1.5 Sintering mechanisms and rates

Mechanism	m	n
Viscous flow	1/2	1
Evaporation–condensation	1/3	$\Delta L/L_0 = 0$
Grain boundary diffusion controlled by vacancy formation	1/4	1/2
Volume diffusion	1/5	2/5
Grain boundary diffusion controlled by vacancy transfer	1/6	1/3
Surface diffusion	1/7	$\Delta L/L_0 = 0$

In this case, in the surface diffusion and evaporation–condensation processes, shrinkage does not occur; therefore, $\Delta L/L \simeq 0$. However, in regular sintering, this ideal situation does not occur; furthermore, since grain growth accompanies sintering, the real phenomena are complicated. Since impurities and additives have large effects, rate manipulation is important, but what is more important is adequate knowledge of changes in microstructure.

Turning to liquid phase sintering, one sees a sintering process facilitated by the intervention of a reactive liquid. This process is used for materials which are difficult to sinter, such as Si_3N_4 and SiC, and temperature reducing materials, as well as cermet-like materials used for reinforcement. In this case, the wetting of the solid particles by the liquid at high temperatures and the solubility of the solid in the liquid are very important. The WC–Co sintered alloy and TiC–NrMo cermet systems are well-known examples.

1.11 WHAT IS HOT PRESSING?

Standard pressure sintering, which uses the surface stress originating in surface energy to effect densification, is commonly used as a means for sintering ceramics. However, surface stress cannot always be relied on, so there is a method using the aid of outside pressure as well as high temperature. This sintering method is called 'hot pressing.' Since in a broad sense all methods of sintering under pressure could be included in this category, ultra-high-pressure sintering and HIP might seem to belong; however, hot pressing is generally understood to be limited to the uniaxial pressure method. In more concrete terms, mold assemblies such as that shown schematically in Fig. 1.18 are commonly employed. A preformed powder compact is placed in the mold, and the sintered product is made by applying pressure at a fixed temperature. Owing to the mechanisms mentioned above, the following are possible: (1) lowering of the sintering temperature; (2) raising of the sintering rate; and (3) densification of difficult to sinter material. As a result, densification is accomplished in a temperature range where exaggerated grain growth and recrystallization do not occur. Thus, a sintered product with satisfactory strength composed of high density small grains can be obtained. In Fig. 1.19, one example of the effects of hot pressing on the densification of ceramics is shown, and in other cases similar effects are obtained.

One very important point in hot pressing is the choice of materials for the die. The die materials shown in Table 1.6 are commonly used, but it is difficult to obtain a long life, inexpensive die material, and there are difficulties with automatic high-speed production. Therefore, use of this method is limited. The most widely used die material is graphite, but depending on the final product, alumina and silicon carbide are sometimes used. Recently a fiber-reinforced graphite system, a thin die of which can take pressures of 300 to 500 kg/cm^2 has been developed.

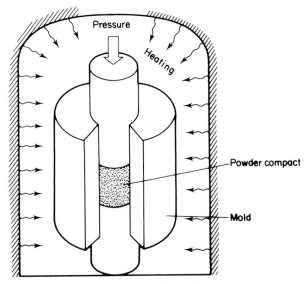

Fig. 1.18 Hot pressing schematic

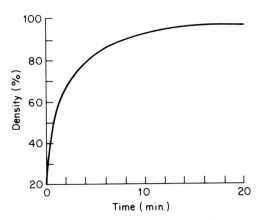

Fig. 1.19 MgO powder compact hot pressing densification example (1,300°C, 280 kg/cm²)

An example of the use of hot pressing is silicon nitride. Silicon nitride powder is mixed with magnesium or another sintering accelerator, and densification takes place at 1,700 °C under a pressure of 300 kg/cm². Since, in this case, the silicon nitride will react with graphite and form a silicon carbide surface, the graphite die is coated with boron nitride, which serves as a mold releasing agent as well as a reaction inhibitor.

The use of a mold releasing agent is normally considered necessary in hot pressing. The stress that occurs during cooling, which is due to the difference in expansion coefficients of the die and product materials, must also be taken into account. And it has been confirmed that high-strength sintered products are obtained by the hot pressing of a Si_3N_4–Y_2O_3–Al_2O_3 system.

Table 1.6 Hot pressing (uniaxial) die materials

	Maximum temperature (°C)	Maximum pressure (kg/cm²)	Comments
Graphite	2,500	700	Neutral atmosphere
Alumina	1,200	2,100 ⎫	Difficult to machine
Zirconia	1,180	⎪	Handling problematic
		⎬	Heat impact susceptibility
Beryllia	1,000	1,050 ⎭	Creep susceptibility
SiC	1,500	2,800 ⎫	Difficult to machine
TaC	1,700	560 ⎬	Reactive
WC, TiC	1,400	700 ⎭	Expensive
TiB₂	1,200	1,050 ⎫	Expensive, difficult to machine
W	1,500	2,450 ⎬	Oxidation susceptibility
Mo	1,100	210 ⎭	Creep susceptibility
Inconel ⎫			
Hastalloy ⎬	1,100		Creep susceptibility
Stainless steel			
Steel ⎭			

Some examples of the application of hot pressing in industrial production include alumina ferrites, boron carbide, and boron nitrate. The silicon nitride mentioned above and silicon carbide, etc. are examples of engineering ceramics still in the development stage. Since boron nitride is a soft ceramics, it is easy to manufacture and has a cost advantage over other materials.

The high cost of hot pressing may be seen in terms of additive value, but the development of a low-cost method is expected for the production of standard industrial products.

1.12 WHAT IS HIP?

HIP is an acronym for 'hot isostatic press.' In sum, this is a technique in which gas pressures of 1,000 to 3,000 kg/cm² and temperatures ranging from several hundred to 2,000 °C are used to compress the product. HIP is a new technique that was developed in the 1950s at Battelle (USA), introduced in industry in the 1960s, and introduced in Japanese industries for alloy manufacture in the 1970s.

An example of a HIP system is shown schematically in Fig. 1.20. Since a high-pressure/high-temperature environment must be created, the materials and structure of the heater and pressure chamber make up a special device.

Pressure sintering of powder, production of high density sintered products, removal of casting defects, diffusion joining, etc. are examples of technological applications of HIP. The first two are applied to ceramics and are called the capsule method and the capsule-free method.

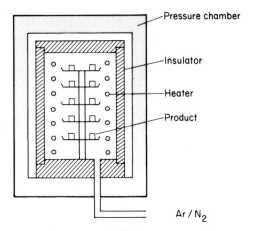

Fig. 1.20 Structure of HIP apparatus

(1) Capsule method

The practical application of this method to ceramic sintering developed out of research into the densification of difficult-to-sinter materials like silicon nitride (Si_3N_4) and silicon carbide (SiC). With this method, additives, which can be considered disadvantageous to high-temperature materials, are reduced in quantity as much as possible, but compared to hot pressing, complicated shapes are possible, and there is the effect of producing isotropic moldings. In the cases of Si_3N_4 and SiC, common glass is used for the capsule material, and the glass is removed at the end of the manufacturing process, but it is important to make sure that no reaction takes place between the product and the glass.

Near 2,000 °C, the exact measurement of temperature is also an important technology. In Fig. 1.21, the pressure/densification figures for Si_3N_4 are shown in graphic form.

(2) Capsule-free method

Normally, sintered products have some percentage of pores, and micro-structural defects exist; therefore, the mechanical chacteristics vary widely and poor products occur easily. This tendency is especially observable in large-scale or difficult-to-sinter materials. To get a product over 90% dense, the intragranular spaces must be blocked; therefore, only the HIP process is effective and produces completely dense sintering.

Use of this method has made possible the removal of residual pores in superhard tools and in ferrite production, and the results of the application of HIP to Si_3N_4 and SiC are available. An example of the results of the use of HIP with Mn–Zn ferrites is shown in Fig. 1.22.

The effects of using HIP are as follows: (1) improvement of mechanical characteristics such as strength and toughness, and decrease in product variation; (2) lowering of sintering temperature; and (3) easy control of grain size.

The capsule and capsule-free methods have been mentioned above as being used for ceramic sintering, but they are batch processes only; therefore, whether production is cost-efficient or not is the biggest concern in determining their wide-scale use.

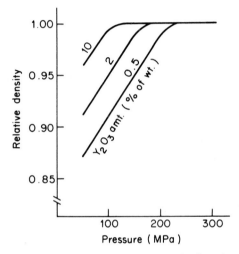

Fig. 1.21 Si_3N_4 glass capsule sintering

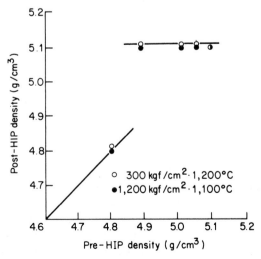

Fig. 1.22 Density changes before and after Mn–Zn ferrite capsule-free HIP

1.13 WHAT IS CVD?

CVD stands for 'chemical vapor deposition,' a high-temperature gas phase reaction. Examples are thermal decomposition, hydrogen reduction, and high-temperature gas reduction of metal halides, in which the metals first and then the oxides and other inorganic materials are precipitated out. These techniques were developed as a means of coating, and though they have continued to be used for coating refractories, they have recently been used for the purification of high-purity metals, powder synthesis, and thin film semiconductor manufacture. They have formed a unique technological field. The distinctive characteristics of CVD are: (i) synthesis of materials with high melting points can be done at low temperatures; (ii) there are many types of precipitates such as single crystal, polycrystal, fiber, powder, film, etc.; (iii) the coating, not only of substrate surfaces, but also of powders is possible. Low temperature synthesis, especially, contributes to energy conservation, which is one of the major aims of modern technology. Examples are αAl_2O_3 and SiC which can be synthesized around 1,000 °C, but for which a low temperature synthesis has become possible.

Roughly speaking, CVD processes may be divided into two categories: the gas reduction of a metal halide with carbon, nitrogen, boron, etc., and the thermal decomposition of a gas on the heated surface of a substrate. Fig. 1.23 shows a schematic CVD device, comprising a vaporization and carrier gas purification chamber, a reaction chamber and a waste gas chamber. Recently new equipment along this line is being developed with mass production in view.

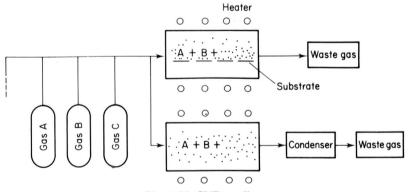

Fig. 1.23 CVD outline

CVD is carried out in a multicomponent gas phase system which includes base gases and by-product gases which vary according to the reaction, and carrier gases. Correspondingly, during the coating process, a diffusion layer forms in the area between the heated substrate and the gas flow. This layer formation can be said to be the primary factor affecting the densification of the coating. A rough illustration of a diffusion layer is given in Fig. 1.24.

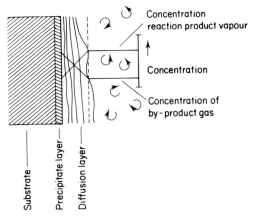

Fig. 1.24 Formation of diffusion layer between substrate and gas flow

Because of this diffusion layer, which is composed of many types of molecules, the precipitation process is complicated. In powder synthesis, seed formation and growth control are the most important factors.

The following processes are part of the new CVD technology: (i) layer flow CVD, (ii) liquid bed, (iii) pyrolytic spray, (iv) plasma CVD, and (v) vacuum CVD. With (i), coated particles may be formed as shown in Fig. 1.25. (An example is SiC, C coated UO_2.) With (iv), a precipitate can be produced at lower temperatures. These and other CVD possibilities are being widely developed.

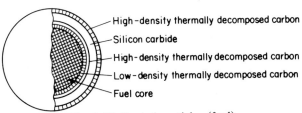

Fig. 1.25 Coated particles (fuel)

1.14 WHAT KINDS OF FUNCTIONAL CERAMICS ARE THERE? IN WHAT FIELDS ARE THEY USED?

Originally, ceramics have various properties, from which a specific functionality is developed. These ceramics are made to exhibit normally unobtainable, unique characteristics. These functional ceramics are called 'fine ceramics.' As is shown in Table 1.7, the functions of ceramics are divided as follows: electric and electronic, magnetic, optical, chemical, thermal, mechanical, and biological. High-strength materials such as silicon nitrate (Si_3N_4) and silicon carbide (SiC), which have become topics for discussion recently, are generally added to the category of fine ceramics with mechanical functionality.

Table 1.7

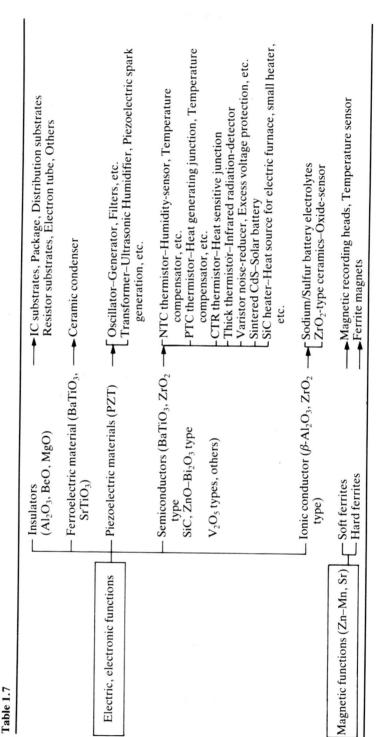

Electric, electronic functions

Insulators (Al₂O₃, BeO, MgO) → IC substrates, Package, Distribution substrates, Resistor substrates, Electron tube, Others

Ferroelectric material (BaTiO₃, SrTiO₃) → Ceramic condenser

Piezoelectric materials (PZT) → Oscillator–Generator, Filters, etc. / Transformer–Ultrasonic Humidifier, Piezoelectric spark generation, etc.

Semiconductors (BaTiO₃, ZrO₂ type, SiC, ZnO–Bi₂O₃ type, V₂O₅ types, others) → NTC thermistor–Humidity-sensor, Temperature compensator, etc. / PTC thermistor–Heat generating junction, Temperature compensator, etc. / CTR thermistor–Heat sensitive junction / Thick thermistor–Infrared radiation-detector / Varistor noise-reducer, Excess voltage protection, etc. / Sintered CdS–Solar battery / SiC heater–Heat source for electric furnace, small heater, etc.

Ionic conductor (β-Al₂O₃, ZrO₂ type) → Sodium/Sulfur battery electrolytes / ZrO₂-type ceramics–Oxide-sensor

Magnetic functions (Zn–Mn, Sr)

Soft ferrites → Magnetic recording heads, Temperature sensor
Hard ferrites → Ferrite magnets

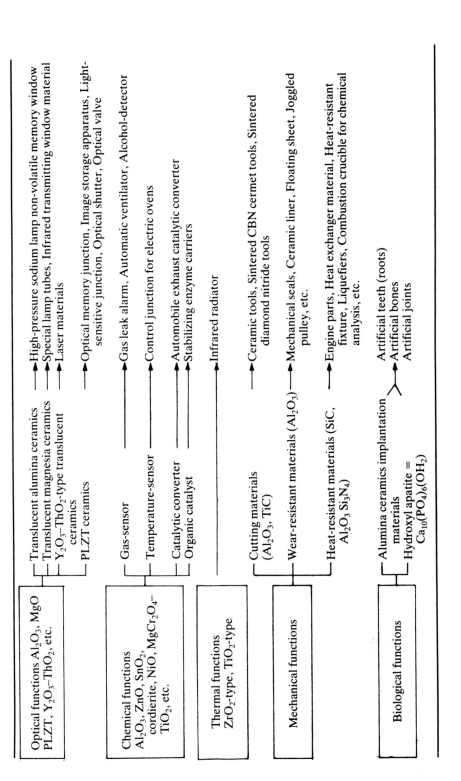

Optical functions Al₂O₃, MgO PLZT, Y₂O₃–ThO₂, etc.

Translucent alumina ceramics → High-pressure sodium lamp non-volatile memory window
Translucent magnesia ceramics → Special lamp tubes, Infrared transmitting window material
Y₂O₃–ThO₂-type translucent ceramics → Laser materials

PLZT ceramics → Optical memory junction, Image storage apparatus, Light-sensitive junction, Optical shutter, Optical valve

Chemical functions Al₂O₃, ZnO, SnO₂, cordierite, NiO, MgCr₂O₄–TiO₂, etc.

Gas-sensor → Gas leak alarm, Automatic ventilator, Alcohol-detector

Temperature-sensor → Control junction for electric ovens

Catalytic converter → Automobile exhaust catalytic converter
Organic catalyst → Stabilizing enzyme carriers

Thermal functions ZrO₂-type, TiO₂-type → Infrared radiator

Mechanical functions

Cutting materials (Al₂O₃, TiC) → Ceramic tools, Sintered CBN cermet tools, Sintered diamond nitride tools

Wear-resistant materials (Al₂O₃) → Mechanical seals, Ceramic liner, Floating sheet, Joggled pulley, etc.

Heat-resistant materials (SiC, Al₂O₃ Si₃N₄) → Engine parts, Heat exchanger material, Heat-resistant fixture, Liquefiers, Combustion crucible for chemical analysis, etc.

Biological functions

Alumina ceramics implantation materials
Hydroxyl apatite = Ca₁₀(PO₄)₆(OH₂)

→ Artificial teeth (roots)
Artificial bones
Artificial joints

(1) Electric and electronic materials

The materials in these fields are insulators, ferroelectric materials, piezoelectric materials, semiconductors, and ionic conductors, which are made from alumina (Al_2O_3), barium titanate ($BaTiO_3$), lead titanate–lead zirconate systems ($PbTiO_3$–$PbZrO_3$), zinc oxide systems (ZnO–Bi_2O_3), β-alumina, etc. Their applications are shown in Table 1.7. The ceramics in this field are the main ones discussed in Section 1.15, on electronic ceramics.

(2) Magnetic materials

Ferrites have this type of functionality, and they are divided into soft and hard types. Among the soft ferrites are the spinel (e.g. $NiFe_2O_4$) and garnet (e.g. $Y_3Fe_5O_{12}$) types, and among the hard ferrites is the magnetic plumbite ($BaFe_{12}O_{12}$) type. Their practical applications are shown in Table 1.7.

(3) Optical materials

Typical examples of translucent ceramics are alumina (Al_2O_3), magnesia (MgO) and yttria (Y_2O_3). PLZT is one of the better-known translucent piezoelectric ceramics. Applications of the better-known types are noted in Table 1.7.

(4) Chemically functional materials

In this field there are ceramic sensors such as gas sensors, humidity sensors, etc., as well as catalysts. There are also oxides such as tin oxide (SnO_2) and zinc oxide (ZnO), and composite oxides (e.g. $MgCr_2O_4$–TiO_2), etc. The practical applications cover a wide range.

(5) Thermal materials

Zirconia (ZrO_2) and titanium oxide (TiO_2) are used for infrared ray emission, which is used as a heat source.

(6) Biologically functional materials

These are listed in Table 1.7, and each has its own field of application. Ceramics with biological applications are discussed in detail in Section 5.1.

1.15 WHAT KINDS OF ELECTRONIC CERAMICS ARE THERE? WHAT ARE THEIR APPLICATIONS?

The materials used in the electronics industry because of their electric and magnetic characteristics are called electronic ceramics. In electronic ceramics, surfaces, grain boundaries and bulk structures are precisely controlled, and this field was the first of the high additive value ceramics fields. Electronic ceramics are used widely in the areas of energy, electronics, home electricity, industrial electricity, automobiles, and other products. An electronic ceramics tree is shown in Fig. 1.26, and functions, uses, and representative materials are laid out in Table 1.8. What follows is a look at some of the representative materials.

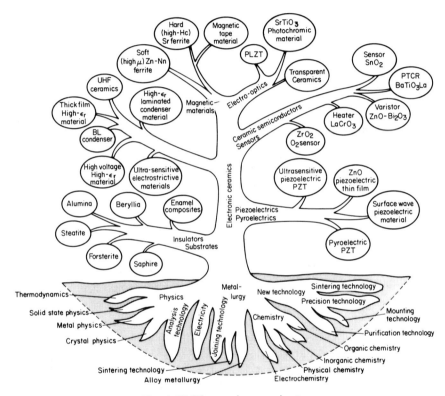

Fig. 1.26 Electronic ceramics tree

(1) Ceramic substrate materials

Insulators hold the most important position in electronic ceramics. Specifically, hybrid integrated circuit insulators and their package materials, which must be made accurately on the micrometer level, use a high-purity, precisely measured, dense-sintered alumina. High-purity dense alumina combines an insulating property not found in metals with a heat-conducting property not found in high polymer materials.

Table 1.8

Function	Use	Materials
Insulation	IC substrates	Al_2O_3, $MgAl_2O_4$, BeO
Electric conductivity	Resistive heat generators, Solar electric generators/Electrodes	ZrO_2, $MoSi_2$, $LaCrO_3$, SiC, Tl, Zr, Hf-borides
Piezoelectronics	Ignition junction	Lighter (automatic lighter) $Pb(Zr, Ti)O_3$
	Piezoelectric filters / Surface wave devices	FM/TV	ZnO, $LiNbO_3$
	Piezoelectric oscillators	Clocks	Quartz, $LiNbO_3$
Semiconductivity	Thermistor	Thermometers	Fe-Co-Mn, Si-O, $BaTiO_3$
	Wireless semiconductors	Varistor	ZnO-Bi_2O_3, SiC
	Gas-absorbing semiconductors	Gas-sensor	SnO_2, ZnO
Magnetism	Hard magnets	Ferrite magnets	$(Ba, Sr)O \cdot 6Fe_2O_3$
	Soft magnets	Transformers, Memory junctions	$(Zn, M)Fe_2O_4$ (M:Mn, Co, Ni, Mg), Garnet-type ferrites
Inductance	Low frequency use	$BaTiO_3 + SnO_2 + Bi_2O_3$
Ionic electric conduction	Batteries	β-Al_2O_3, ZrO_2 $(+CaO, Y_2O_3)$

(2) Ceramic piezoelectric materials

Mutual exchange between an electric signal and a mechanical signal is carried out through a piezoelectric junction. Piezoelectric junctions shaped to requirements are mainly manufactured by sintering $PbTiO_3$ or $PbZrO_3$ (PZT) type systems. A piezoelectric effect is also obtained in the polarization of single crystals in sintered substances.

The main uses of piezoelectricity are ignition coils and resonators. Resonators work as filters that allow only a chosen fixed frequency of radio waves to pass and are essential junctions in the tuning circuits of television sets and radios.

(3) Ceramic semiconductor materials

There are also many types of ceramic semiconductors. Thermistors, in which resistance changes with temperature are in common use. The resistance of a NTC thermistor decreases with increasing temperature, demonstrating a general characteristic of semiconductors. Iron group metal ceramic oxides are widely used for thermistors which are used for temperature control, because they exhibit chemical and thermal stability and are easy to manufacture. Thus what are called PTC thermistors are manufactured from a $BaTiO_3$ ceramics that has been made semiconductive. Since the resistance of this ceramics increases abruptly near the phase transition temperature, it can be used as a heat-sensitive resistance junction, and it will automatically maintain the temperature near the phase transition temperature.

Sintered ZnO in which the grain boundaries or interstices are impregnated with Bi_2O_3, which has a low melting point and high resistance, is used for varistors since it exhibits high resistance up to a fixed voltage, called the varistor voltage, and low resistance above it. Varistors have practical applications as circuit protection junctions and lightning shunting junctions.

1.16   WHAT KINDS OF HEAT-RESISTANT CERAMICS ARE THERE?

Considering the conditions in which high temperature materials are used, one can see that silicon nitrate (Si_3N_4) and silicon carbide (SiC), etc., which are compounds with extremely strong covalent bonds, maintain their high bonding strength at high temperatures. Furthermore, their thermal expansion coefficients are low, and their corrosion resistance is excellent. Therefore they are the best materials for high temperature structural parts. In Table 1.9, the data supporting this conclusion for representative ceramics, Young's modulus (E), density, and thermal expansion coefficient are given. When the theoretical strength σ_{th} is equal to $E/10$, Si_3N_4, SiC, AlN, etc., in which E is large, provide high bonding strength, while the thermal stress caused by internal heat distribution at high temperatures is low. Thus their suitability as high-temperature materials is understandable. However, because the strength of the material is determined by the cracks that exist within the structure, with the exception of whisker and filament structures which are close to the ideal case, only a strength of the order of 1/100 of the theoretical value can be obtained. This is explained by the fact that the local stress at the point of the crack just before fracture has the same value as the bonding strength of the material.

Table 1.9 Basic characteristics of various ceramics

Material	Density (g/cm³)	Elasticity (kg/mm²)	Melting point/ decomposition temperature (°C)	Heat expansion coefficient (10⁻⁸ degC⁻¹)
AlN	3.26	3.4×10^4	2,450	4.9
Al_2O_3	3.99	3.6×10^4	2,050	8
BeO	3.02	3.8×10^4	2,530	10
SiC	3.25	5.7×10^4	2,600	4.3
Si_3N_4	3.2	3.8×10^4	1,900	2.5–3
Quartz glass	~ 2.2	0.7×10^4	—	0.6

The fact that Si_3N_4 and SiC, etc. are stable and show little loss of strength at high temperatures means that they are difficult to sinter. The material planning assessment problems are the densification of these materials and the building up of process engineering that arises from the functional level requirements for high-strength/high-temperature ceramics.

Five main conditions necessary for ceramics that maintain high strength at high temperatures are as follows: (i) the compound must have strong covalent bonds (e.g. Si_3N_4, SiC, and AlN, etc.); (ii) the actual density must be near the theoretical density; (iii) grain size must be small and have uniform distribution; (iv) the shape of particles must be anisotropic (plate-like, needle-like, etc.); (v) the grain boundary phase between the particles must have high heat-resistance. Si_3N_4, SiC, and AlN are essentially high-bonding-strength materials of the type mentioned above, but this is if one is speaking in ideal terms. In reality one must be satisfied with a number of peculiar conditions in these polycrystalline materials.

A completely reliable knowhow has not been established for high-temperature/high-strength ceramics, but the characterization of the process from the raw materials to the final product is the main point under investigation. But beyond the mechanical techniques, which are basic, the techniques for monitoring the many factors which affect strength during manufacture do not differ from other processes. Fig. 1.27 shows the factors governing the strength of ceramics from the points of view of physical properties, microstructure, and external effects.

This fundamental technology will help research development and effect the extension of high-temperature/high-strength ceramics, and will help to enhance the performance of these ceramics.

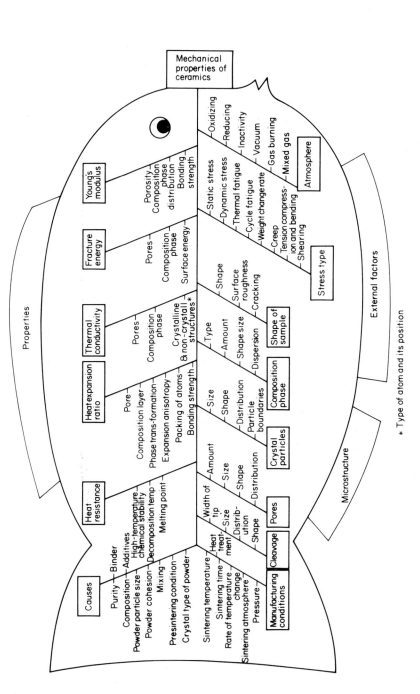

Fig. 1.27 Factors controlling ceramic strength

* Type of atom and its position

1.17 WHY DO CERAMICS TRANSMIT LIGHT?

If a material is to be transparent, it must allow light to pass through. A beam of light incident to the surface of a uniform material is either reflected by the surface or absorbed by the material with a resulting dissipation of energy. Any remaining light shows up as transmitted light. Correspondingly, it is desirable to have poor reflection and absorption characteristics in translucent materials. However, reflection is determined by the difference between the refractive indices of air and the material, and this, being a relative concept, is not a determining factor with respect to transparency.

On the other hand, absorption is an essential element in determining the transparency of a substance. Light absorption is either intrinsic absorption, accompanying an absorption of energy, or a dispersion with no true absorption. Intrinsic absorption is due to an absorption of light energy through an excitation of electrons within the substance and is an inherent property determined primarily by the kinds of atoms and their structure. Correspondingly, a necessary condition for transparency in a substance is that there be no inherent absorption in the visible light spectrum (0.4–0.7 μm).

With dispersion, there is no energy absorption by the material, and there is a light-scattering phenomenon. In general, as mentioned in Section 1.5, the microstructure of ceramics consists of crystalline grains, grain boundaries, precipitates, and pores. The three origins of the scattering that occurs when light is propagated in an asymmetric single-system medium are shown in Fig. 1.28. They are: (i) scattering caused by the pores left by the sintering process, phase irregularities in the distribution and precipitation of additives, and mixed composition within a single phase; (ii) scattering caused by aggregations of imperfections, such as empty lattice points, and dislocations in the crystalline structure, which can be regarded as grain boundary scattering; (iii) light scattering (optical irregularity), etc., caused by optical anisotropy in microcrystals and reflection and refraction (multiple refraction) in discontinuous grain boundary interfaces.

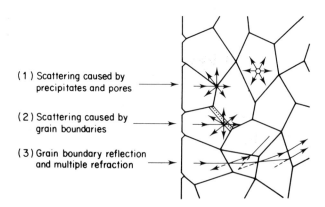

(1) Scattering caused by precipitates and pores

(2) Scattering caused by grain boundaries

(3) Grain boundary reflection and multiple refraction

Fig. 1.28 Main light scattering effects inside transparent ceramics

Of these three processes, the optical irregularities in optically anisotropic grain boundaries are inherent properties of the material, but the others are related to ceramic sintering processes. Accordingly, since ceramics that transmit light must not absorb light, it is necessary that they be materials that do not exhibit optical anisotropy. A cubic system of crystallization is preferable in order to avoid this anisotropy. In addition, even though alumina (Al_2O_3) and beryllia (BeO) have hexagonal crystallization systems, they are optically uniaxial, and the light that permeates the structure is transmitted by way of scattering. It is even possible for a uniaxial or biaxial optically anisotropic substance to transmit light if the magnitude of the refractions is low, and the substance is close to being isotropic.

Through control of the primary causes of scattering such as precipitates and pores, in effect control of the grain boundaries, a number of ceramics, shown in Table 1.10 are possible light-transmitting materials. Here one must distinguish between 'translucent' and 'transparent.' For substances mentioned so far in which there is a scattering effect, 'translucent' is used, and for substances like glass, in which there is no scattering, only absorption, 'transparent' is used.

Table 1.10

Composition	Crystal structure	Melting point (°C)	Hot pressing (°C)	(kg/cm²)	Sintering (°C)
			Manufacturing example		
Al_2O_3	Hexagonal	2,050	1,500	400	1,950
BeO	Hexagonal	2,570	1,200	2,000	1,800
MgO	Cubic	2,800	770	100	1,750
CaO	Cubic	2,570	1,150	600	
Y_2O_3	Cubic	2,410	900	800	2,200
ZrO_2	Cubic	>2,700	1,300–1,750	500–30,000	1,450
ThO_2	Cubic	3,300			2,100
$MgO \cdot Al_2O_3$	Cubic	2,135			2,150
CaF_2	Cubic	1,360	900	2,600	
GaAs	Cubic	1,240	900–1,000	600–3,000	
PLZT		1,450	1,000–1,300	200–600	

2 Structural Ceramics: Questions and Answers

Chapter 2 is concerned with structural ceramics. First, the kinds of oxide and non-oxide ceramics, and their characteristics are explained. Next, alumina (Al_2O_3) and zirconia (ZrO_2), which are representative oxide ceramics, are taken up. Nitrides and carbides are the core of non-oxide ceramics, so silicon nitride (Si_3N_4), aluminum nitride (AlN), boron nitride (BN) and silicon carbide (SiC), all of which are commonly discussed, are taken up. Finally, the ceramics used in nuclear fuel systems are touched upon.

2.1 HOW ARE OXIDE CERAMICS CLASSIFIED AS STRUCTURAL MATERIALS? WHAT ARE THEIR CHARACTERISTICS?

Before being developed into high-performance products, ceramics were well-known as the products of brickmaking, glassmaking and pottery-making. After all, pottery, china, glass and cement, which are casually encountered everyday, can be called structural ceramics here. These are called traditional ceramics, and they have a long history. Therefore, although the raw materials from which their constituents are made are either natural materials like clay, or not so strictly controlled hand-worked materials, they achieve their functional purpose, and one can roughly grasp the process. Of course glass cement, refractories, etc. are not produced by such simple methods nowadays, and as a result of technical control from the raw materials to the final products they have become the basis for modern industry.

What can be said about general characteristics? A brief sketch follows. First, these materials are composed largely of elements like Si, Al, Mg, Ca, Na, etc., and of course O because they are oxides; therefore, they are found in great quantities on earth. This point is very important. Large amounts of cement and glass are used for buildings and roads, but when it is realized that

43

44

there is an almost inexhaustible supply, their continued consumption can be understood. Since their history is long, they are products of a field with high-level technical accomplishments. This is the same as saying that it would not be a mistake to acknowledge that these Japanese products are on the highest level in world terms. Recently, the Japanese have been apt to be dazzled by fields that present new technical challenges, but the importance of these long-established ones may be much greater.

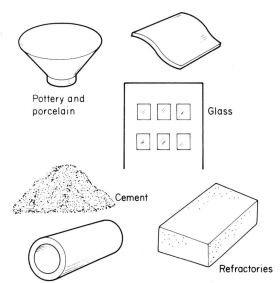

Fig. 2.1 Structural oxide ceramics

In pottery and porcelains there have recently been introduced many varieties and the compositions have become complex. But the basics are that as the firing temperature increases, they become more dense, harder and more transparent. Firing temperatures usually range from about 900 °C to 1,500 °C. Also the general tendency is for the inherent colors to disappear.

Refractories, which are used in high-temperature industrial processes, melt at high temperatures and do not change shape easily. They are used in electrical and various other types of furnaces, but when used their heat resistance must be carefully analyzed. Of course, the atmosphere inside the furnace controls their usage life. Correspondingly, there are various bases for classification, including chemical properties, heat resistance, usage, shape, etc.

Glass is amorphous; that is in general terms, it is a non-crystalline solid in which the molecular configuration has no fixed order. Classifications cover many fields, and depending on the constituents, properties and usages, there is a complicated nomenclature. But the same representative categories mentioned above and many others apply.

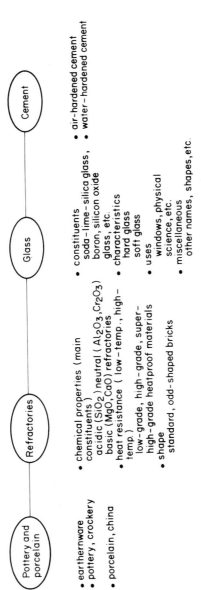

Fig. 2.2 Classifications of structural oxide ceramics

Pottery and porcelain
- earthernware
- pottery, crockery
- porcelain, china

Refractories
- chemical properties (main constituents)
  acidic (SiO_2) neutral (Al_2O_3, Cr_2O_3)
  basic (MgO, CaO) refractories
- heat resistance (low-temp., high-temp.)
  low-grade, high-grade, super-high-grade heatproof materials
- shape
  standard, odd-shaped bricks

Glass
- constituents
  soda-lime-silica glass, boron, silicon oxide glass, etc.
- characteristics
  hard glass
  soft glass
- uses
  windows, physical science, etc.
- miscellaneous
  other names, shapes, etc.

Cement
- air-hardened cement
- water-hardened cement

The fundamental characteristic of cement is that it solidifies at standard temperatures through a reaction with water. Once again there are various types with various uses. As indicated above, in order to get satisfactory results from the use of these many types and functions, careful consideration is needed.

As mentioned above, the basic materials employed in civil engineering and used in kilns and furnaces, what can be called structural materials, affect daily life in ways that do not call attention to themselves. Naturally, recent technical innovations extend to these fields. And since these fields involve high-temperature processes that consume much energy, the advanced development of energy conservation in production processes is particularly striking.

2.2 HOW ARE NON-OXIDE CERAMICS CLASSIFIED AS STRUCTURAL MATERIALS? WHAT ARE THEIR CHARACTERISTICS?

Non-oxide ceramics is a general term for the ceramics made from metal carbides, nitrides, sulfides, silicides, borides, etc. In recent years, the characteristics required of materials cover a broad range. In the field of structural materials, especially in the area of heat-resistant and high-temperature structural materials, the user conditions are such that existing oxide ceramics will not hold up. But there is hope that materials will become more resistant. Also, this is the direction in which the key materials to a new technical breakthrough lie. This chapter is limited to structural materials, but non-oxides are especially attractive as materials with new functions.

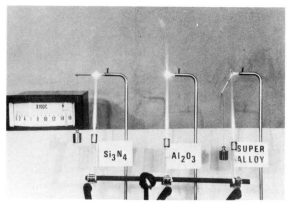

Fig. 2.3 High-temperature strength comparison of non-oxide ceramics and other materials

The question is, 'What kinds of new structural materials can be expected?' and the main items in the list would have the characteristics of lightness and resistance to deformation even at high temperatures, and corrosion and wear resistance. Also the availability of abundant, cheap resources is a condition that should not be overlooked. The background of these conditions is a series of social needs such as conservation of natural resources and energy, progress toward high efficiency heat engines, and the limits to and search for replacements for scarce resources. Carbides and nitrides are the most prominent non-oxide ceramics, and because the degree of covalency in their bonds is large, one can predict that they will show a strong resistance to deformation., Both SiC and Si_3N_4 are thought to satisfy the necessary conditions noted above. Worldwide study of these materials began between 1970 and 1972.

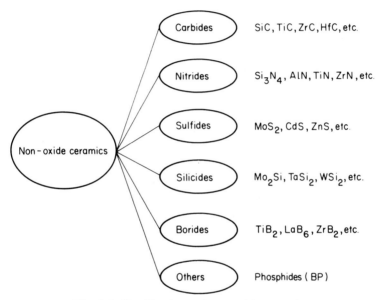

Fig. 2.4 Classifications of non-oxide ceramics

However, it is difficult to arrive at something that can be regarded as an industrial material through the mere extension of existing materials engineering, and it is natural that new engineering tasks will be produced. A one-by-one explanation of these materials will be found in the separate sections that follow, but briefly, the one problem they have in common can be stated as 'the progress of the reliability of ceramics.' Advances in processes and evaluation methods are indispensable to this progress, and the people involved are making every effort to achieve them. This one point is expected to make a large contribution to the general advance of ceramic engineering.

With these tendencies as a basis, there has been a lot of discussion about how materials should be made. In effect, from basic material synthesis all the way up to reliability evaluation technology, a new system of material science is coming into being. There is great potential for non-oxide ceramics in various areas which have usage conditions under which metals will not hold up, but it is very difficult to realize the inherent superior functionality of these materials in concrete form. Yet one must be strong enough to accept the challenge of bringing out new materials.

2.3 WHAT KINDS OF CERAMIC NITRIDES ARE THERE?

Nitrides are compounds of nitrogen expressed by the formula Me_xN_y (Me: metal), and along with carbides, which will be discussed later, they have been noted in recent years, among special ceramics as materials which cover the areas in which oxide ceramics are weak. It is well-known that there are very many kinds of nitrides, but all of them are man-made minerals that are not found in nature. Despite the many kinds, those with material possibilities are limited. That is to say, boron nitride (BN), aluminum nitride (AlN), silicon nitride (Si_3N_4) and titanium nitride (TiN) are the main ones.

The general properties are as follows: crystal structures are mostly hexagonal or cubic, and densities cover a wide range from about 2.5 to 16 g/cm³ (eg., nitrides of Ta or W); also in a restricted atmosphere (non-oxidizing), heat resistance is very high. These give one a rough conception. If one's thinking is limited to structural materials, the classifications according to fields of usage are: (1) high-strength machine parts, (2) corrosion- and wear-resistant parts, (3) heat-resistant parts, and (4) other structural maintenance parts. Thus, when normal metals or oxide ceramics are not quite satisfactory, or when a rather far-fetched idea has to be realized, one can say these are materials to look to for interesting results. The position of these materials within industrial materials as a whole has not solidified, and since it is settling out, the other side is that a great deal of data needs to be accumulated.

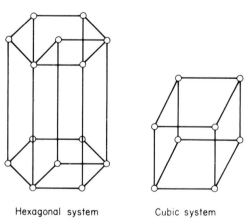

Hexagonal system   Cubic system

Fig. 2.5 Typical crystal structures

Table 2.1 displays the properties of representative nitrides, but these are only the basic properties, and for other properties related to ceramics, refer to the sections on the individual ceramics.

Table 2.1 Typical nitride properties

Substance	Melting point (°C)	Density (g/cm^3)	Resistivity (Ω cm)	Thermal conductivity (cal/cm^2/cm/s °C)	Thermal expansion coefficient (× 10^6 degC^{-1})	Hardness (Mohs)
HfN	3,310	14.0	—	0.0517	—	8–9
TaN	3,100	14.1	135 × 10^{-6}	—	—	8
ZrN	2,980	7.32	13.6 × 10^{-6}	0.033	6–7	8–9
TiN	2,950	5.43	21.7 × 10^{-6}	0.070	9.3	8–9
ScN	2,650	4.21	—	—	—	—
UN	2,650	13.52	—	—	—	—
ThN	2,630	11.5	—	—	—	—
Th$_3$N$_4$	2,360	—	—	—	—	—
NbN	2,050 (decomposition)	7.3	200 × 10^{-6}	0.009	—	8
VN	2,030	6.04	85.9 × 10^{-6}	0.027	—	9
CrN	1,500 (decomposition)	6.1	—	0.021	—	—
BN	3,000 (sublimation/decomposition)	2.27	10^{13}	0.036–0.069	0.59–10.51	2
AlN	2,450 (sublimation/decomposition)	3.26	2 × 10^{11}	0.048–0.072	4.03–6.09	7–8
Be$_3$N$_2$	2,200 (sublimation/decomposition)	—	—	—	—	—
Si$_3$N$_4$	1,900	3.44	10^{13}	0.004–0.005	2.5	9

Here are a few comments on the properties. First, extremely high melting points are characteristic. However, many of these materials are inclined to decompose at atmospheric pressure, and it is a mistake to think they will last up to the stated temperature. In practice, if the data for the real material is not investigated carefully, there is a chance that estimates will be faulty. As for resistance, a wide range is covered from insulating materials to conductors. In low-resistance materials, the crystal structure is one in which nitrogen is added to the original cubic (etc.) structure of the metal, and the original structure is retained; thus the resistance is close to that of the original metal. The materials in which the crystal structure is different from the original turn into insulators.

2.4 WHAT KINDS OF CERAMIC CARBIDES ARE THERE?

Carbides are a group of compounds expressed by the general formula Me_xC_y (Me: metal), which includes materials with extreme hardness and extremely high melting points. Because of these characteristics, these are important materials for engineering ceramics. However, carbides, like nitrides, are limited because exposure to an oxidizing atmosphere causes unavoidable oxidation to CO_2 and a metal oxide. Silicon carbide is an exception and has a fairly good oxidation resistance. SiC, boron carbide (B_4C), titanium carbide (TiC), zirconium carbide (ZrC), vanadium carbide (VC), tantalum carbide (TaC), tungsten carbide (WC), molydenum carbide (Mo_2C), etc., can be given as representative carbides.

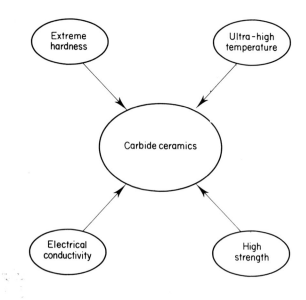

Fig. 2.6 Characteristics of carbide ceramics

SiC is a well-known and widely used industrial material, and both TiC and WC are used for cutting tool tips and wear-resistant coatings. The extreme hardness of B_4C gives rise, for example, to its use in mortars, and its light weight gives rise to its use in bulletproof vests and the control rods for nuclear reactors. In addition, uranium carbide (UC) has applications as a nuclear fuel.

The chemical compositions of carbides are known to have wide stoichiometric limits. Thus Me_2C changes to MeC_2 and for each of a number of other carbides there are several compositions. But it is characteristic of the carbides of Si, Ti, Zr and Hf to have only one. Of these, both HfC and TaC have high melting points of the order of 3,900 °C, and thus have possibilities as light bulb filaments.

A great difference between these compounds and nitrides is their electric conductivity; and use of this property in SiC can be seen in the heating elements of high-temperature furnaces. And since ceramic carbides have this kind of heat resistance, extreme hardness accompanied by wear resistance, and electric conductivity, they can be used as functional structural materials.

In Table 2.2, a number of characteristics which are typical of SiC, TiC, WC and TaC are shown. From these, the general nature of these compounds can be deduced. It can be seen that melting points and hardness values are extremely high. With one exception, electrical resistance is low, and the compounds are good conductors. SiC is generally known to be a semiconductor, but in ultra-pure, single-crystal form, it is understood to show insulating characteristics. However, obtaining such high purity in an industrial material is technically and economically difficult, and it is a problem to be faced in the near future. With continued interest in the synthesis of high-purity materials, the further expansion of the new functionality can be expected.

The examples in this section are limited, but the same things can be said about other groups of carbides. It is important that the needs of material synthesis and the requirements of practical use come together. Here the difference between a substance and a useful material must be understood. Even though a substance has great potential, whether or not it can become a useful material depends on manufacturing processes, planning, design, etc. The 'if it doesn't work the first time, it's no good' way of thinking seems common, but plunging another step forward is recommended.

2.5 ALUMINA (Al_2O_3) IS A TYPICAL CERAMIC MATERIAL. WHAT IS IT?

Lately ceramics have been used in diverse fields. Alumina (Al_2O_3) is typical, and can be considered a fine ceramics. The properties relevant to alumina ceramics' use as a structural material are, briefly, an extremely high melting point (2,050 °C), hardness, electrical insulation, and the ability to take on diverse shapes and functions.

Table 2.2 Properties of representative carbides

Properties	SiC	TiC	WC	TaC
Density (g/cm^3)	3.2	4.8	15.8	14.5
Melting point (°C)	~3,000	~3,100	~2,700	~3,900
Hardness (kg/mm^2)	~2,500	~3,000	~2,000	~2,000
Specific heat capacity (cal/g °C)	0.16 (RT)	0.2	0.056 (265 °C)	0.19 (RT)
Thermal conductivity (cal/cm s °C)	0.08–1.2	0.08–1.1		
Thermal expansion coefficient (degC^{-1})	$4\text{–}5 \times 10^{-6}$ (RT ~ 2,000 °C)	7.2×10^{-6} (RT ~ 800 °C)	2.9×10^{-6} (RT ~ 2,200 °C)	6.7×10^{-6} (RT ~ 1,900 °C)
Young's modulus (kg/mm^2)	$3.5\text{–}7.0 \times 10^4$	4.6×10^4	6.9×10^4	3.7×10^4
Poisson's ratio	0.19	0.19		
Electric resistance (Ω cm)	$10^{-5}\text{–}10^{+13}$	$\sim 10^{-4}$	$\sim 10^{-6}$	$10^{-4}\text{–}10^{-5}$
Crystallization system	cubic, hexagonal	cubic	hexagonal, cubic	cubic

Table 2.3 Properties of alumina ceramics

Property	Alumina
Water absorption (%)	0–0.00
Apparent relative density	3.4–3.7
Thermal expansion coefficient 25–700 °C ($\times 10^{-4}$)	7.5–7.9
Tensile strength (kg/cm²)	1,400–1,750
Compression strength (kg/cm²)	10,000–28,000
Bending strength (kg/cm²)	2,800–4,200
Wear strength (kg cm/cm²)	5.6–6.2
Thermal conductivity (cal cm/cm² s °C)	0.040–0.045
Insulating capacity (V/mm)	10,000
Volume-based resistance 100 °C	2.0×10^{13}
300 °C	5.0×10^{10}–6.0×10^{11}
500 °C	1.0×10^{8}–1.0×10^{9}
700 °C	3.0×10^{6}–4.0×10^{7}
T_c value (°C)	800–930
Dielectric constant　　　1 mC/s	8.3–9.3
10,000 mC/s	8.0–9.1
Dielectric loss ($\times 10^{-4}$) 1 mC/s	3–7
10,000 mC/s	14–15

Toshiba's High-Density Hybrid Modules

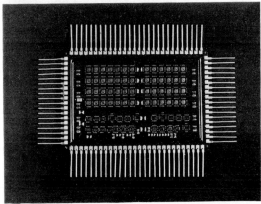

Fig. 2.7 High-density mounting substrate

The raw materials of alumina should be mentioned here. Alumina (Al_2O_3), which is the oxide of aluminum (Al), is second only to silica (SiO_2) among the natural minerals in the amounts that exist in nature. A large proportion of it exists in feldspar, mica and other alumino-silicates, and pure Al_2O_3 is only found in small amounts. Bauxite, which is well-known, was given the name of the place in France where it was discovered in 1821. It is hydrated Al_2O_3, and is composed of a mixture of $Al_2O_3 \cdot H_2O$ and $Al_3O_2 \cdot 3H_2O$. Subsequent research made clear what is known today, that hydrated alumina exists in diaspore (α-$Al_2O_3 \cdot 3H_2O$), beohmite (γ-$Al_2O_3 \cdot H_2O$), hydrargillite (γ-$Al_2O_3 \cdot 3H_2O$) and bayerite (γ-$Al_2O_3 \cdot 3H_2O$) forms. Correspondingly, bauxite is a general term for the ores of aluminum hydroxides. There are small amounts of corundum (anhydrous alumina) found in nature. Al_2O_3 can also be obtained artificially from various chemical salts through chemical treatment by either the wet method or the dry method. In addition, fused alumina which is in a stable crystal state is produced by fusing simple raw materials and is composed of α-Al_2O_3 crystals, which are the same as natural corundum.

Another important thing to know about alumina is its transformations. Beside α-Al_2O_3, there are γ, δ, θ, β, ξ, etc., polymorphic varieties. The cubic spinel structured configuration has cation lattice vacancies. At temperatures over 1,000 °C, it is transformed into the α configuration, but the reverse does not occur. Also, when the β configuration shows up in $Na_2O \cdot 11Al_2O_3$, it is known as a sodium ion conductor.

Now, where is alumina used? Recently it has held an important position in the so-called mainstay industries and is a key material in IC and ultra-LSI (large-scale integration) semiconductor substrates, thermal head substrates for facsimile machines, transparent alumina sheaths for high-voltage sodium lamps (known under the name 'Lucalox'), automobile parts and sparkplugs. This shows that the insulation, high thermal conductivity, high strength and anti-corrosion properties of alumina give rise to plenty of products. Other uses of alumina include artificial rubies, ceramic tools (bits) which make use of its wear resistance, cermet (e.g. Al_2O_3–Cr) systems, and abrasives.

2.6 WHAT IS ZIRCONIA?

Zirconia is an oxide described by the formula ZrO_2. It is a new fine ceramics, and its development from now on will be worthy of notice. Because it has an extremely high melting point of about 2,700 °C, its original potential was as a refractory material; but its use would be limited because of its peculiar transformations. These transformations proceed in three stages:

$$\text{Monoclinic} \xrightarrow{\text{1,000-900 °C}} \text{tetragonal} \xrightarrow{\text{2,370 °C}} \text{cubic}$$

The transformations are accompanied by great changes in volume, and if the material is used without modification, cracks or other consequential changes arise. However, the addition of Y_2O_3, CaO, etc., into the crystal structure is known to maintain the highest temperature configuration (cubic) even at low temperatures. This is called 'stabilized zirconia.' This is an important material for solid electrolytes, since ZrO_2 has highly ionic bonds and it has a fluorite (CaF_2) type, cubic crystal structure. The fluorite structure is a face-centered cubic lattice, with anions in the center of each of the eight tetrahedrons it forms and a large open space within the inner octahedron. Thus, the diffusion of the anions, which form a simple cube, is easy compared to that of the cations. Besides which, anion vacancies exist in greater abundance that would normally be created by additives and temperature conditions. This discussion

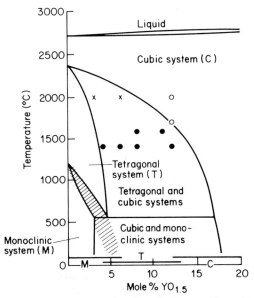

Fig. 2.8 ZrO_2–Y_2O_3 system configuration (H. G. Scott "Phase Relationship in the Zirconia-Yttria System," *J. Mat. Sci.*, **10** 1527 (1975).)

ends here with the general concepts, and details are left to books about crystal chemistry.

There is also a partially stabilized zirconia, better known by its acronym, PSZ. The martensite-like transformation of tetragonal ZrO_2 ↔ monoclinic ZrO_2, shows the effect of absorbing the energy produced when the ZrO_2 configuration is breaking apart. Because of this, though ceramics are usually fragile, a difficult-to-break, high-toughness ceramics can be made. In effect, the absorption of the energy suppresses the progress of molecular collapse. This is called 'stress-induced phase transformation,' and its development is noteworthy in terms of the use of high-toughness ceramics as structural materials. The structure of PSZ is as follows: as noted before, ZrO_2 with a stable cubic configuration is synthesized at high temperatures by the addition of CaO or Y_2O_3. Next, if it is annealed at temperatures at which the tetragonal configuration is stable, minute tetragonal crystals will be precipitated, and a mixture of the cubic and tetragonal configurations will be produced. Thus, the mixture is partially stabilized by the part that is in the cubic configuration. The solid solution with Ca(Y) is expressed by the formula $(Zr, Ca(Y))O_{2-x}$, because the migration of O^{2-} ions is possible since oxygen lattice vacancies are produced. Because of this solid electrolyte capability, it is used on oxygen sensors (oxygen concentration cells). Also there are high expectations for PSZ in high-temperature machine parts, because of its high toughness.

This toughness will open the way to uses that do not exist now, such as ceramic scissors and knives; hence the development of this material will be exciting.

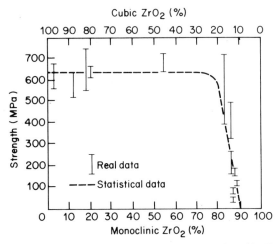

Fig. 2.9 Strength vs cubic system ZrO_2 ceramics composition (T. K. Gupta Rate of stress-induced phase transformation in enhancing strength and toughness of zirconia ceramics, Fracture Mechanics of Ceramics, 1978, Plenum Press. (pp 877–889)

Table 2.4 Toughness of various materials

Materials	K_c (MN m$^{-3/2}$)
ZrO_2 (semi-stable)	6–9
ZrO_2 (stable)	1.1
Si_3N_4	4.8–5.8
SiC	3.4
B_4C	6.0
Al_2O_3	4.5
Sapphire (single-crystal)	2.1
Spinel (single-crystal)	1.3

Source: T.K. Gupta, *J.Mat.Sci.*, **9**, (1974).

One problem with PSZ is that semi-stable configurations exist in equilibrium at high temperatures, so that even in the temperature range in which the change to the stable configuration should occur, PSZ is not always formed. At present improvements in the high-temperature characteristics are being studied.

2.7 WHAT ARE THE CHARACTERISTICS AND APPLICATIONS OF CERAMIC SILICON NITRIDE?

In recent years, silicon nitride (Si_3N_4) has drawn much attention as a high-temperature industrial material, and has developed in the last 10 years to the point that it ranks with silicon carbide and the oxides. This ceramics which was first developed in Great Britain in the 1960s and energetically studied in the United States, Japan and West Germany in the 1970s, is now a main current high-temperature, high-strength ceramics. To state some of the properties of Si_3N_4, it is light (relative density 3.19 g/cm³), hard (Vickers hardness 1,900 g/mm²), strong (Young's modulus 3×10^4 kg/mm²), and has a low thermal expansion coefficient (3×10^{-6}). The reasons that this ceramics can be considered for engine materials and other mechanical parts that up to now only metals could be considered for, that is for parts that are used at ultra-high temperatures, are as follows. Si_3N_4 differs from oxides in that it does not 'melt' at high temperatures; rather, decomposition and sublimination into Si and N between 1,800 °C and 1,900 °C is characteristic, and it is resistant to deformation at high temperatures because its molecular bonds are rich in covalence. It would be great if the seemingly possible ceramic materials could be extracted from the characteristics of Si_3N_4. The problem is how to realize them.

Table 2.5 Typical properties of Si_3N_4 ceramics

	Reaction-sintered Si_3N_4	Hot pressed Si_3N_4	Standard pressure sintered Si_3N_4
Maker		Norton	GTE Sylvania
Density (g/cm³)	2.4–2.6	3.2	3.24
Thermal expansion coefficient (degC⁻¹)	2.5–3.2×10^{-6} (25–1,000 °C)	3.2×10^{-6} (20–1,000 °C)	3.4×10^{-6} (25–1,000 °C)
Thermal conductivity (kcal/m h °C)	3.5 (RT)		24 (RT) 12 (1,000 °C)
Specific heat capacity (kcal/kg °C)		0.26 (RT)	0.19 (RT) 0.28 (1,100 °C)
Young's modulus (kg/mm²)	1.5–1.82×10^4 (RT)	3.2×10^4 (RT)	2.8×10^4 (RT)
Poisson's ratio			0.23 (RT)
Bending strength (kg/mm²)	30 (RT) 35 (1,200 °C)	101 (RT) 77 (1,100 °C) 49 (1,315 °C)	67 (RT) 52 (1,200 °C) 29 (1,400 °C)
Tensile strength (kg/mm²)		42 (RT) 28 (1,100 °C) 18 (1,315 °C)	
Hardness (kg/mm²)			1,300 (Knoop)
Quantitative increase with oxidation (mg/cm²)	10 (1,400 °C–20H)	0.7–0.8 (1,400 °C–20H)	

First the sintering process is carried out. Generally sintering accelerators are added to the powdered raw material, and it is then thermally densified at temperatures above 1,700 °C. There is another method in which densification takes place without the use of sintering accelerators (reaction-sintering). This shows the production of the ceramic material, but says nothing about the extremely important engineering involved in the processes that shape it into usable parts. This means that plans for objects that are of practical use must be satisfied. Performance must be verified by stress analysis, prototype manufacture, parts testing, and practical testing. Also, parallel materials development and development of evaluation techniques are necessary for the development of high-temperature materials. In the near future, the systematization of materials science will also put these to use.

Next, let us look at the fields in which Si_3N_4 may be used. Usage in high temperature machine parts centers on a wide range of areas related to energy, and even if restricted to high temperatures, high strength, light weight and high corrosion-resistance, there are many areas of application, as is shown in Table 2.6. In particular, the development of efficient heat engines for cars, which is perpetuated by the future increasing dependence on petroleum, will necessarily turn to ceramics. In any event, the development of this material can be thought of as being in the region of unexplored technology, and though the hurdle that must be cleared is a high one, it can be cleared.

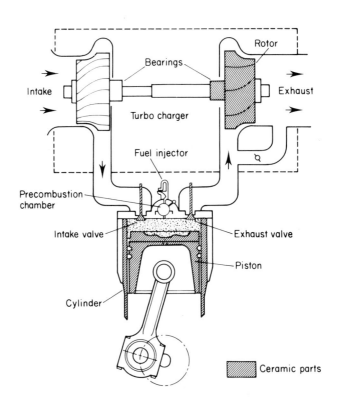

Fig. 2.10 Conception of a ceramic diesel engine

Table 2.6 Uses of Si_3N_4 ceramics

Field	Part name		
	Stationary blade	Actual type	
		Radial type	
Gas turbines	Stationary blade Burners Nose cone shroud		
Diesel engines	Piston cap Cylinder Exhaust valve head		
Other heat engines	Nuclear fusion lining material Heat pump		
Heat-, corrosion-resistant jigs and tools	Materials for use with molten metals Bearings, mechanical seals Bearings Processing tools		

2.8 WHAT ARE THE NEW TECHNIQUES FOR CERAMIC SILICON CARBIDES?

Silicon carbide has been ranked with silicon nitride as a high-temperature non-oxide ceramics since the 1970s. SiC ceramics has been known for a long time and has been put to practical use. SiC is synthesized by either carbon reduction of SiO_2 or direct carbonization of Si. The typical crystal structures are cubic (β) and cubic (α), but SiC exhibits much polymorphism, and single-crystal synthesis is being studied. The crystal periods vary slightly according to the conditions of synthesis, etc.

So, how has it been a topic in ceramics over the past 10 years? It first began with a paper published by Dr Drochazcha, a General Electric researcher. It was noted that the covalent bonds of SiC are extremely strong and carrying the sintering far enough is very difficult; however, mixing in small amounts (~1%) of boron (B) and carbon (C) and sintering around 2,000 °C makes sintering to the theoretical density possible. This meant a great deal from the point of view of sintering. Whether SiC sintering is liquid-phase sintering or solid-state sintering is not always clear, but a sintering process in which no liquid is discernible was produced. This may be explained by the fact that the thermodynamic conditions are satisfied because boron and carbon cause a change in the surface conditions of SiC powder. Until then, sintering via a large volume liquid phase had been common knowledge, but the occurrence of a liquidless sintering with a small amount of additives became a subject for worldwide attention. Since then a number of additives have been studied, but generally B centered carbon systems, for example B_4C, are known to be effective. Because of this, desirable SiC ceramics can be obtained with standard pressure sintering.

There is a reaction-sintered SiC ceramics that is made using the reaction, Si + C→SiC. There are no additives, but in return, there is a comparatively large number of pores; thus, there is the fault of low material strength. Furthermore, SiC has been made by soaking SiC in fused Si.

Anticorrosive, wear-resistant machine parts and high-temperature gas turbine parts are thought of as the main applications for SiC ceramics. Compared with Si_3N_4, one can expect greater corrosion resistance in this type of machine part. Up to now, low strength has been a difficulty, but with a little improvement, its prospects can be considered very good.

On the other hand, the physical properties of SiC are worth noting. Compared with other ceramics, its thermal conductivity is extremely high. Generally, SiC has impurities and is semiconductive, but researchers at Hitachi have said that if it is sintered with the addition of BeO, a small amount forms a solid solution, and it becomes an insulator. It has the thermal conductivity of Al and the insulating property of Al_2O_3, and is notable as an ideal packaging material for small-size integrated semiconductors.

Fig. 2.11 SiC machine parts (Toshiba Ceramics, Inc.)

Table 2.7 Properties of SiC ceramics

	Standard pressure sintered SiC	CVD SiC	Reaction-sintered SiC
Density (g/cm³)	3.2	2.9	3.2
Thermal expansion coefficient (degC⁻¹)	4.8 (RT ~ 1,500 °C)	4.8	4.4 (RT ~ 1,000 °C)
Thermal conductivity (cal/cm s °C)	0.2 (RT)	0.2 (RT)	0.2 (RT)
Specific heat capacity (kcal/kg °C)	0.1	0.3 (1,200 °C)	—
Young's modulus (kg/mm²)	4.1×10^4	4.9	4.1
Bending strength (kg/mm²)	35–43 (RT ~ 1,600 °C)	~ 50 (1,200 °C)	46–44 (RT ~ 1,000 °C)
Hardness (kg/mm²)	2,800	3,000–4,000	2,800

2.9 WHAT IS SIALON?

More than four years have passed since the study of the development of silicon nitride (Si_3N_4) as a high-temperature, high-strength material began, and during this time, new substances and materials have been found. Among them Sialon has attracted the greatest amount of attention. The word originates as follows. When Si_3N_4 with Al_2O_3 added is sintered, it is known that the Al_2O_3 forms a solid solution in the Si_3N_4, and the series of materials formed is known by the names of the compositional elements, Si–Al–O–N. This is 'Sialon.' 'Sialon' is used as a noun, and it holds a place among the many Si_3N_4 system–oxide group materials.

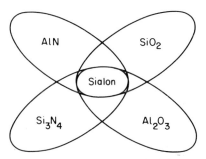

Fig. 2.12 Composition of Sialon

The properties of Sialon are (i) comparatively little expansion (2–$3 \times 10^{-6} \deg C^{-1}$) and (ii) high corrosion resistance (it has the quality of reacting with an oxidizing atmosphere at high temperatures, and for example, the deterioration of strength is comparatively greater than that of other Si_3N_4 system materials), but since the characterization is not complete, one, unfortunately, cannot draw any conclusions. Therefore, only a partial description can be given here.

First, how is Sialon synthesized? If a Si_3N_4–Al_2O_3–AlN–SiO_2 system (four elements) is reacted at 1,700 °C to 1,800 °C, Sialon is produced. Many kinds of Sialon are synthesized here. This is not a simple reaction, and the conditions are such that other phases are created before the main reaction is complete. Among them, the solid solution mentioned above is called 'β-Sialon,' and thus has the same configuration as β-Si_3N_4. (There are two types of Si_3N_4, α and β.) This solid solution is expressed by the formula

$$Si_{6-x}Al_xO_xN_{8-x} \ (O \times 4.2)$$

As shown in Fig. 2.13, there are other types of Sialon which are different from β-Sialon, and new fields of oxide chemistry and crystallography are being opened up. It is rare to obtain a single configuration when synthesizing new materials. This is because the reactions take place among many ingredients and are dominated by time factors and the factors of the basic material synthesizing processes. Sometimes, β-Sialon alone is called Sialon, and the two characteristics mentioned above may be considered typical.

62

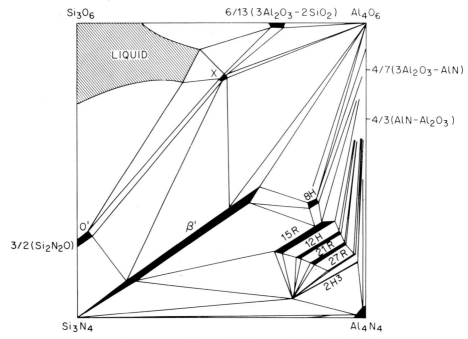

Fig. 2.13 Configuration of Sialon (K.H. Jack, *J. Mat. Sci*, **11**, 1135 (1976))

The wide range of applications that is being considered includes engine parts, corrosion-resistant jigs, etc. This is repetitive, but once again, the synthesis of single phases and complete control are the central technical problems, and truly useful material must await the results of upcoming research. Naturally, it will take a little while to characterize the other substances in Fig. 2.13. Thus, the applications and science of Sialon, with its many undeveloped fields, are being researched on the primary level.

2.10 WHAT ARE THE CHARACTERISTICS AND APPLICATIONS OF ALUMINUM NITRIDE (AlN)?

Nitride ceramics are new high-temperature materials worthy of notice. Among these, aluminum nitride (AlN) is a new ceramics that is expected to be useful as an industrial refractory material, because wetting by molten metals does not easily take place; it is highly stable in a non-oxidizing atmosphere at high temperatures in the area of 2,000 °C, and its thermal impact resistance is comparatively high. Although it was called 'new,' AlN has a long history. Synthesis began with Geuther in 1896, and at the beginning of the twentieth century, it became a well-known intermediate for nitrogen fixation in the Serpek method. This is a method in which a powder mixture of bauxite and coke is heated to around 1,750 °C in an atmosphere containing nitrogen. AlN is produced in a reaction that follows the formula $Al_2O_3 + N_2 + 3C \rightarrow 2AlN + 3CO$. This AlN is then reacted with water to synthesize ammonia (NH$_3$). Use of this method ceased with the development of the later Harber method, but since it shows such physical and chemical possibilities, investigations of industrial uses continue. Not only is there interest in it as a structural material, but also research into the synthesis of single-crystal AlN, which is a III–V family compound, for electronics use as well as thin film and many other kinds of research, is being carried out.

The methods for producing AlN are heating of a mixture of Al_2O_3 and C powder in a N_2 or NH_3 (Serpek method) atmosphere, the direct reaction of Al and N_2 or NH_3, the reaction of Al halides with NH_3, etc. In order to produce a useful material, a sintered product must be made, and there are hot pressing, standard pressure sintering, reaction-sintering and other methods available. A high-density sintered product may be obtained by hot pressing under pressure at 1,800 to 2,000 °C. AlN is a rather highly ionic compound for a non-oxide; correspondingly, material transfer proceeds independently to some extent, and a driving force such as the external pressure of hot pressing allows for densification. Standard pressure sintering is carried out at approximately 1,800 °C with the use of sintering additives (Y_2O_3, Al_2O_3, SiO_2, BeB, CaO, etc.). Reaction-sintering is a method for obtaining dense products by the nitrization of Al. Even if the heat production of the Al–AlN system nitrization is controlled, a certain amount of pores will remain, and only simple sintered products can be obtained, because the reaction is remarkably exothermic.

Fig. 2.14 Sintered AlN

Practical applications of AlN include non-melting material for molten metal crucibles, protection tubes, vacuum vaporization chambers (such as those for gold (Au), because of its low vapor pressure in a vacuum), refractory bricks, refractory jigs, etc. The refractory bricks are ideal for use in 2,000 °C class non-oxidizing electric furnaces, because their thermal resistance in special atmospheres is excellent. Another use for AlN is in heat radiation plates. Because its thermal conductivity is 2 to 3 times that of Al_2O_3, and its hot pressed products are stronger than Al_2O_3, it is used in fields with requirements for high thermal conductivity under high stress, such as the insulation of semiconductor junctions and heat radiation substrates in motor vehicles. The combination of thermal conductivity, strength, and heat resistance make AlN a very worthwhile substance.

Table 2.8 Properties of aluminum nitride ceramics

Property	Normal sintering		Hot pressing	
	AlN	AlN–Y$_2$O$_3$	AlN	AlN–Y$_2$O$_3$
Density (g/cm^3)	2.61–2.93	3.26–3.50	≈ 3.20	3.26–3.50
Porosity (%)	10–20	≃ 0	2	≃ 0
Color	grey-white	black	black-grey	black
Fracture strength (kg/mm^2)	10–30	45–65	30–40	50–90
Hardness (kg/mm^2)	—	1,200–1,600	1,200	1,200–1,600
Elasticity coefficient ($\times 10^4$ kg/mm^2)	—	3.10	3.51	2.79
Thermal expansion coefficient (degC^{-1}) 25–1,000 °C	5.70	—	5.64	4.90
Heat conductivity (cal/cm s °C) 200 °C	—	—	0.07	—
800 °C	—	—	0.05	—
Machinability	good	good	good	fair
Oxidation resistance	poor	excellent	good	excellent

2.11   BRIEFLY EXPLAIN BORON NITRIDE (BN)

Boron nitride is called white carbon, since it closely resembles carbon, and amorphous, hexagonal and cubic crystal systems exist. As shown in Fig. 2.15, hexagonal boron nitride resembles graphite, and boron and nitride form a hexagonal net-like structure with the layers linked by weak van der Waals bonds. But it is distinguished from graphite because it exhibits an electrical insulation property and white color. Therefore this material is used not only for heat- and corrosion-resistant products and electrical insulators, but also as a B diffusion material for semiconductors. On the other hand, cubic boron nitride has a diamond-like structure, so it is used as a cutting tool material.

In general, hexagonal boron nitride is formed by hot pressing, and the sintered product is machined. In this way, since boron nitride can be effectively processed and machined, precision ceramic parts may be produced easily. In contrast, cubic boron nitride is synthesized like diamond by applying ultra-high pressure to a fine particle powder compact.

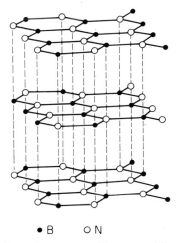

• B   O N

Fig. 2.15 Crystal structure of hexagonal boron nitride

Next, the characteristics of these boron nitrides can be given. The hexagonal configuration has high heat resistance, thermal decomposition at 2,200 °C, and high corrosion resistance. It is easily machined, and is an excellent lubricant and electrical insulator. On the other hand, cubic boron nitride, like diamond, has extreme hardness and high thermal conductivity, etc. The fundamental characteristics of boron nitride and its products are summarized in Tables 2.9 and 2.10 respectively.

There is abundance and variety in the practical applications of boron nitride which include heat-resistant, anticorrosive materials such as metal melting pots, jigs, single-wheel transfer pumps, high-temperature furnace materials, boats, molds for glass, etc.; electrical materials such as insulators, infrared and microwave light polarizers and permeable materials, ICs, transistors, heat-radiation panels for microwave modules, and semiconductor boron diffusion sources; and lubricants in various solid forms, powders, and sprays. It is also used in nuclear reactor shields and pressure transmission intermediates. The above are the general uses of formed products and powders, but special cubic boron nitride exhibits great power in tool and grinding materials.

Table 2.9 Basic properties of hexagonal boron nitride

Density (g/cm^3)	2.3 (hexagonal system)
	3.5 (cubic system)
Melting point (°C)	3,100–3,300 (N$_2$ 30,000 atmospheric pressure)
	2,700 (decomposition, 1 atmospheric pressure)
Specific heat capacity (C_p) (cal/g °C)	0.2–0.5 (RT800 °C)
Thermal expansion coefficient (degC^{-1})	-2–-3×10^{-6} (a-axis)
	20–23 $\times 10^{-6}$ (c-axis)
Heat conductivity (W/cm °C)	13 (crystallographic estimation)

Table 2.10 Properties of hexagonal BN formation products

Property	Pyrolitic BN		Hot pressed BN
	a-axis direction	c-axis direction	
Density (g/cm^3)	~ 2.1		~ 2.1
Strength (kg/mm^2)	4.2 (RT)	—	6–11 (RT)
	10.1 (2,200 °C)	—	
Heat conductivity (cal/cm s °C)	0.15 (RT)	0.004 (RT)	0.04–0.07
	0.15 (815 °C)	0.007 (815 °C)	
Thermal expansion coefficient (degC^{-1})	-2×10^{-6}	20×10^{-6}	0.8–7.5 $\times 10^{-6}$
Electric resistance (Ω cm)	3×10^7 (980 °C)	3×10^9 (980 °C)	2×10^{14} (RT)

2.12 WHAT ARE THE CHARACTERISTICS OF TITANIUM CARBIDE AND TUNGSTEN CARBIDE?

There are many carbides of transition elements such as titanium carbide (TiC), tungsten carbide (WC), tantalum carbide (TaC), zirconium carbide (ZrC), etc., all of which have similar characteristics. They are characterized by extreme hardness and wear resistance. Of these, TiC and WC are widely used for industrial products. Both of them, with the addition of metals are used in the form of cermets, when the latter is called a 'super-hard alloy.'

(1) Titanium carbide cermets

Since the latter half of the 1950s, these cermets have been put to use in high-speed finishing tools for steel. Before that, ceramic tools that were made mainly of alumina had been used, but there was a problem with brittleness. As shown in Fig. 2.16, these cermets were developed as tool materials that can be placed as intermediates between ceramics and super-hard alloys. They are manufactured with TiC as the main constituent by adding Ni and Mo and sintering. TiC-type cermets show superior anti-pitting wear-resistance characteristics, and if they are used in the same speed range as super-hard alloys, they have a longer usage life. Other aspects include prettier finishing surfaces, and favorability over WC in terms of resources.

Based on these characteristics, they are used for high-speed steel finishing cuts and throw-away tips, but are not suitable for discontinuous cutting or rough processing.

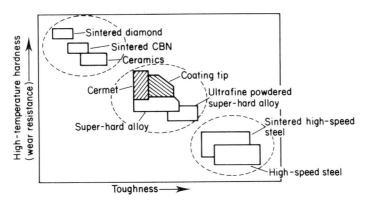

Fig. 2.16 Classifications of tool materials

(2) Tungsten-carbide-based super-hard alloys

Super-hard alloy has WC as its main constituent, and it has a long history. The production of it in Japan started in 1928. During the Second World War, its use in cutting tool and die materials spread rapidly, and up to this day, it has been used in a wide variety of fields as the main material for tools. In terms of composition, there are two types; the (i) WC–Co system and (ii) WC–TaC–TiC–Co system types. These are made using WC as the main constituent, and after pressure forming, they are sintered. Compared with metal tools, these have superior characteristics in that they have great hardness up to high temperatures, excellent wear resistance, high resistance, high compression strength, and excellent impact resistance. Because of this, they are used not only for cutting steel, cast steel, and steel system materials, but also for cutting cast iron, non-ferrous metals and non-metals. Table 2.11 shows a comparison of the characteristics and uses of various other tool materials.

Table 2.11 Characteristics and uses of blade materials

Blade material	Standards	Characteristics	Major uses
Tool-type carbon steel Tool-type steel alloy	JIS G 4401 JIS G 4404	Inexpensive but high-temperature hardness is low	Slow cutting
Tool-type high-speed steel	JIS G 4403	High-temperature strength and toughness are excellent. Prescribed capacities can be obtained by thermal treatment, but the temperatures change with constituents and proper temperature ranges are narrow. Therefore caution is necessary	WO systems: general cutting W–V systems: cutting of difficult-to-cut materials Mo systems: discontinuous cutting, heavy cutting
Super-hard alloys	JIS H 5501 JIS B 4053	Made from high-melting-point carbides of tungsten, titanium, tantalum (WC, TiC, TaC), etc. with binding materials (Co) by powder metallurgical methods and sintering. Hardness and wear resistance are excellent at both normal and high temperatures	P-type: steel-cutting types M-type: general and difficult-to-cut material cutting K-type: cast materials, non-ferrous metals, etc.
Ultra-fine powder super-hard alloys		Super-hard alloys using ultra-fine particle WC, extreme toughness	Heavy/discontinuous cutting, fine deep cutting, slow cutting
Coated super-hard alloys		Basic materials are super-hard alloys which are given a several micrometre thick coating of TiC, TiN or Al_2O_3. This forms a combination with the toughness of the basic material	General cutting (all types)

Table 2.11 (*contd.*)

Blade material	Standards	Characteristics	Major uses
Cermets	a CIS 015 CIS 016 CIS 017	Main ingredients, TiC, TiN, are combined with iron system metals. Wear and heat resistance are excellent; stable in reactions with metals and deposition characteristics are excellent	High-speed steel finishing cuts
Ceramics	CIS 014	Sintered oxides with Al_2O_3 as the main constituent; extreme hardness at high temperatures, possibilities for high-speed cutting	High-speed casting finishing cuts
Diamond		Has the greatest hardness, correspondingly, wear resistance is excellent. Expensiveness and fragility are weak points	Non-ferrous metal material precision cutting
Sintered diamond		Fine particle diamond crystals are sintered at high temperatures under pressure to produce a polycrystalline conglomeration. Hardness is extremely high at high and low temperatures; chemically inert to substances other than iron	Non-ferrous metal cutting, non-metal material cutting
Sintered CBN		Fine particles of CBN, which is next to diamond in hardness are sintered at high temperatures under pressure; extremely low reactivity with metals, and extremely stable at high temperatures	Super-hard alloy, quenched steel, chilled casting material finishing cuts

a Cemented Carbide Tool Manufacturers Association.

2.13 WHAT KINDS OF CERAMICS ARE THERE FOR NUCLEAR FUEL SYSTEMS?

Nowadays, nuclear power generation in Japan is done mainly by light water nuclear reactors (boiling water, BWR, or pressurized water, PWR, types), in which UO_2 pellets are used for nuclear fuel. The structure of a nuclear fuel assembly is shown in Fig. 2.17. Inside the nuclear fuel rods are pieces of sintered UO_2 which are called 'pellets.' They are called zircaloy, and nuclear fuel rods are made of a zirconium alloy pipe filled with them.

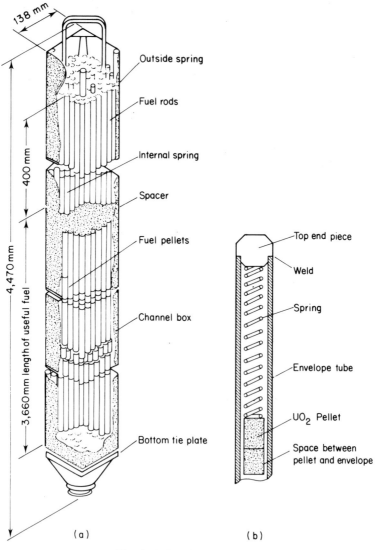

Fig. 2.17 BWR fuel assembly

Here is an explanation of the UO_2 fuel pellets. The main manufacturing process is shown in Fig. 2.18, and this flowchart can be roughly divided into the powder synthesis process and UO_2 pellet sintering process. During operation, the pellets are exposed to high-speed neutrons or nuclear fission fragments, resulting in either a dimensional contraction or the volume increase associated with nuclear fission. In order to prevent these, it is important to control the microstructure of the pellets. Pores are made large, and after sintering, it is necessary that the density be of the order of $94 \pm 1.5\%$. Fig. 2.19 shows the external appearance and the microstructure of the pellets. In the case of a light water reactor, from the point of view of the thermal resistance of the zircaloy envelope tube, it is difficult to raise the internal temperature above 500 °C. Even if stainless steel is used, the temperatures are limited to 550–600 °C. In order to improve this and raise the heat resistance, a particle fuel coated with ceramics has been developed. On

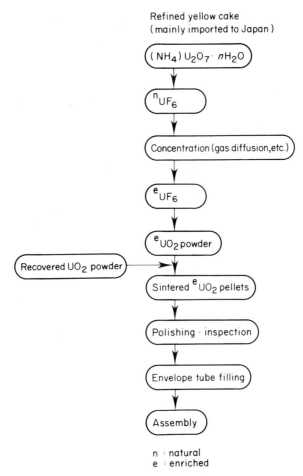

Refined yellow cake
(mainly imported to Japan)

$(NH_4) U_2O_7 \cdot nH_2O$

$^nUF_6$

Concentration (gas diffusion, etc.)

$^eUF_6$

$^eUO_2$ powder

Recovered UO_2 powder

Sintered $^eUO_2$ pellets

Polishing · inspection

Envelope tube filling

Assembly

n : natural
e : enriched

Fig. 2.18 UO_2 pellet manufacturing process

72

Fig. 2.19 UO$_2$ pellet and its structure (JNF Co.). (a) External view of pellet (outside diameter 10.65 mm). (b) Microstructure

the surface, a pyrocarbon, which is impermeable to gases is used, and underneath it a buffer layer coating of SiC, both of which are applied by CVD. This is a noteworthy high temperature nuclear fuel system.

Neutron control rod materials are used to control the output of a nuclear fission reactor. If one takes the BWR as an example, B_4C powder is used as a control rod material, and the combustible poison G_2O_3 is mixed into the UO_2 pellet. In addition, peripheral materials such as decelerators, radiation materials, and shield materials are important. So far, this has been an explanation of the ceramic fuels and the peripheral materials used in light water nuclear reactors, but in the future it is expected that fast breeder reactors and nuclear fusion reactors will be developed. To make these practicable, there are great expectations for a UO_2–PuO_2 system ceramic fuel for the former and a ceramic nuclear lining for the latter. But this will push out the limits of the environments in which ceramics can be used, and the latter, especially, is considered a strong technical barrier.

3 Electronic Ceramics: Questions and Answers

Chapter 3 deals with electronic ceramics. Among the electronic ceramics, magnetic materials such as ferrites and piezoelectric ceramics will be touched upon first. Ferrites are used in radio wave absorbers, magnetic cores for accelerators, etc. There are abundant uses for piezoelectric ceramics in igniting devices, transformers, resonators, etc. Next, the ceramics used in semiconductors will be treated—from various points of view, including uses as temperature, gas, moisture, and other sensors, as well as heating elements and other uses. Uses of optical ceramics, transparent alumina, etc., have become common, so these will also be mentioned.

3.1 CAN CERAMICS ABSORB THE ENERGY OF LIGHTNING?

If lightning hits a power line, there is a high-voltage surge that destroys the insulation in transformers, circuit-breakers, and the like, and results in massive power failure. What protects the power apparatus from this type of surge is a lightning arrester. A lightning arrester consists of a series-wired gap for discharge of electrical surges only and a special element that suppresses voltage rises through its own voltage–current properties. The lightning arrester maintains the normal voltage in the power system with the series gap insulation, acts instantly to suppress the excess voltage in the case of a surge, and suppresses abnormal rises in the system-wide voltage.

This special element that regulates voltage and has current–voltage properties is a ceramics that absorbs the energy of lightning, and the material called SiC is used to make it. This SiC device is made by sintering SiC particles of around 200 μm in size with a binder, and this product shows non-linear current–voltage characteristics because of point contacts among the SiC particles. Since these non-linear current–voltage characteristics are not brought about by any relationship to Ohm's law, $V = IR$, the distinctly

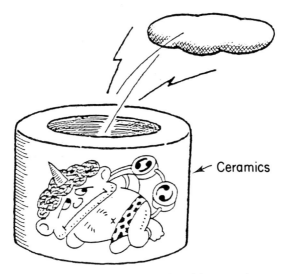

Fig. 3.1 'Lightning' gets trapped by ceramics

non-ohmic voltage–current situation shown in Fig. 3.2 results. Because of this, as mentioned above, when a high voltage is applied across the device, the surge current flows through the device, so that the voltage on both sides of the junction is suppressed to the voltage determined by the resistance of the junction at the time of the maximum current. Recently an alternative to the SiC device has been considered; it is a ZnO device, composed mainly of zinc oxide (ZnO). In this device, 10 μm ZnO crystalline particles are surrounded by approximately 0.1 μm of bismuth oxide or some other thin, high-resistance layer, and these boundary layers form closely adhering three-dimensional contact surfaces that give rise to sharply non-linear current–voltage characteristics. With this non-linearity, a device of 63 mm diameter, 20 mm thick shows an extremely low voltage increase of 1.6-fold with an 8-digit, 10^{-3} to 10^5 A, change in current. Because of this, ZnO devices have an excellent absorption capacity for electrical surges caused by lightning; thus, the device can be small in comparison with the one made of SiC, and a lightning arrester of one-seventh the size is possible.

Even under the condition where the system voltage is applied across the device, an insulating condition can be maintained because there is a low current leakage. There is no necessity for a series gap. In these ways, the ZnO lightning arrester has many excellent properties, but since the raw material, ZnO powder, is used for facial and baby powders, the characteristics of the sintered product are difficult to imagine from the current common conception.

A ZnO lightning arrester is superior to the current ones composed of a series-wired gap and an SiC junction, since the transmission of direct current is possible. The methods available for transmitting electric power are alternating current transmission, which is currently used, and direct current transmission, which is good for long-distance transmission, because of low energy loss during transmission. There are plans for direct current transmission in Japan. A series wired gap cannot be used as an interceptor with direct current, and therefore the emergence of the ZnO lightning arrester could have great significance.

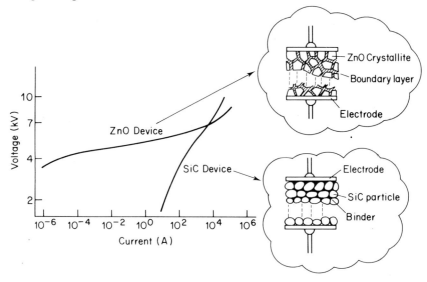

Fig. 3.2 Voltage–current characteristics of the ZnO and SiC junctions

3.2 HOW CAN CERAMICS ABSORB RADIO WAVES

Ferrites are ceramics that can absorb radio waves. Absorption of radio waves is absorption of electromagnetic energy; in other words, it is necessary that the material cause a loss. In general, 'good materials' normally cause little loss of energy, but in this case it is just the opposite and the loss is put to constructive use. There are three types of electromagnetic loss in ceramics: (i) ohmic loss in resistors, (ii) dielectric loss in inductors and (iii) magnetic loss in magnetic substances. Among the substances used for ohmic loss resistors is foamed polystyrene with carbon powder distributed inside it. At 100 MHz, the thickness is about 1.5 m, but its light weight and broad band characteristics make it useful for radio darkrooms. Materials used for their dielectric loss include titanium barium oxide materials which are in trial manufacture, but have not been put into practical use. Ferrites make use of magnetic loss. Ferrites show loss in bands ranging from UHF to SHF, and are worth noting as thin, wide-band radio wave absorbers. They have practical application in the prevention of television ghost lines. With the continued growth of tall buildings in cities, obstacles that reflect the VHF radio waves used for television signals are increasing. As a preventative method, test installations of ferrite radio wave absorbers on the outer walls of buildings have been performed. But the ferrites used have a relative density of 5, and a thickness of 1 cm is necessary; therefore, when 1 m² is put up it means a weight of 50 kg, which is considerable. In consideration of the actual workability, the development of curtain-wall type absorbing walls is proceeding. 10 cm × 10 cm × 1 cm ferrite tiles are set up in a continuous series in the direction of the magnetic field, and vacant space in the direction of the electric field. In the vacant space, bolts and nuts can be arranged, which provides a practical means of installation.

Table 3.1 Characteristics of several types of absorbers

Absorption frequency	Absorber	Absorber thickness
2,000 MHz and below	Ferrite absorber	5–10 mm
From 4,000 to 8,000 MHz	Rubberized ferrite	10 mm or more inversely proportional to frequency
8,000 MHz and above	Carbon rubber	2–3 mm

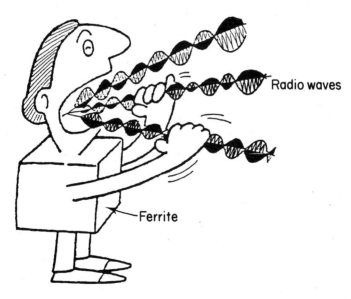

Fig. 3.3 Radio waves being eaten by 'Ferrite'

However, it is regrettable that there is at present no practical success in the use of radio wave absorbers. If a good material with a 10 times greater magnetic loss, a thickness of 1 mm and a weight of 5 kg were found, and if the cost problem were solved, the possibilities for practical use would become greater. Again, though they are not currently in use, there is an example of a study for a practical application. This is the investigation of the use of radio wave absorbers as a countermeasure to the signal jamming that is expected to be caused by the Honshi Bridge, the large bridge planned to connect the Japanese islands of Honshu and Shikoku. With frequencies in the X-band, about 10 GHz, the installation of rubberized ferrite and rubberized carbon in a thickness of about 2 mm to 3 mm was considered possible. Both materials are soft, and it is possible to make large units. However, when the workability problems are considered, one finds that they cannot be applied like paint; therefore, a wait-and-see attitude was taken since there was nothing else with as large a magnetic loss.

But there are many other applications for ferrite radio wave absorbers, other than in construction materials. They have been used for a long time in the attenuators that prevent high-frequency leakage to the direct current power supplies in high-frequency oscillators. Though they do not use ferrites alone, ultra-wide band radio wave matchers that widen the band of absorbed frequencies are made of composites in which a material with conductive loss is coated with ferrite. Also, there are uses for load matching in waveguides.

3.3 IN WHICH PARTS OF A SYNCHROTRON ARE FERRITES USED?

Ferrites are used in the high-frequency acceleration resonators and kicker magnets of proton synchrotrons. The part played by ferrites in high-frequency acceleration resonators is the making of an LC resonator with high resonant impedance that causes the high-frequency high voltage for proton acceleration. Besides this, ferrites to which a magnetic bias field is added make electronic alignments over a wide range of frequencies through changes in L. Kicker magnets are pulse magnets that are used to change the path of the proton beam, and they are used as magnetic cores because they are capable of carrying out high-speed operations.

A proton synchrotron is a giant accelerator for high-energy physics and elementary particle physics research. Proton acceleration in a synchrotron is done by first accelerating the particle in a linear accelerator and then injecting it incidentally into a giant vacuum ring around the inside of which it is accelerated in a magnetic field created by electromagnets. The frequency of revolution is over 1 MHz, and a high frequency high voltage which is synchronized with an integer multiple of this frequency is used. An LC resonator which employs ferrite-loaded coaxial inductance is used to generate this voltage. Since the frequency must be increased along with the acceleration of the proton, the resonator increases its adjusted frequency. Normally the resonator for a high-frequency accelerator is made up of large ferrite rings (outside diameter, 20 cm to 50 cm; inside diameter, 10 cm to

30 cm; thickness, about 2.5 cm), several tens of which are layered coaxially. Figs. 3.4 and 3.5 show their appearance and arrangement.

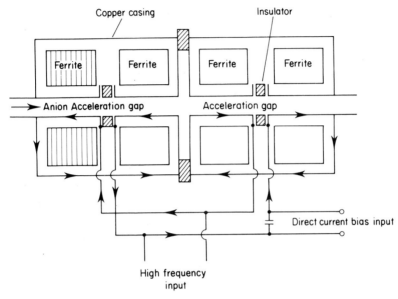

Fig. 3.4 Resonator for a high-frequency accelerator

Fig. 3.5 Ferrite rings

In order to satisfy usage conditions, the ferrite material undergoes various improvements. The quality of the ferrite is determined by the permissible amplitude of magnetic flux density, since the frequency of the acceleration voltage is proportional to the angular frequency, the cross-section of the ferrite core, and the amplitude of the flux density. Furthermore, there is a need to make the resonant impedance as high as possible in order to generate as high an acceleration voltage as possible using the current supplied by the high-frequency output. And it is desirable that the magnetic permeability and the machinability coefficient of the ferrite be large. For satisfying these requirements, NiZn system ferrites are excellent, and in the search to improve the characteristics, ones with Co additives have been developed.

These large-scale ferrites for high-power use are used at the Fermi National Accelerator Laboratory (NAL) in the USA, the High Energy Research Center (KEK) which was built at Tsukuba Academic City in Japan, etc. Proton synchrotrons are installations that require enormous amounts of money for construction and operation, and the ferrites used in high frequency accelerators are extremely important materials for managing the efficiency and performance of these accelerators.

3.4 CAN FERRITES SERVE AS THE MAINSPRINGS OF WATCHES?

There is a magnet which has a strong coercive force and low cost and is called a Ba-ferrite magnet. It is widely used in the magnetic cores of small motors. When a Ba-ferrite magnet is made in the shape shown in Fig. 3.6 and is surrounded by a permalloy core, the magnet moves like a wave. The core has a coil wrapped around it, but when there is no current flowing through it, the core is the same as a simple piece of iron and the magnet is at rest as in Fig. 3.6(a). Next, if a current is connected across the coil, the core becomes an electromagnet and north and south poles are generated in the parts that surround the Ba-ferrite magnet. Since each of the core arms facing the north and south poles of the magnet have changed into matching north and south poles, the repellent force causes a rotation (clockwise) toward the opposite poles as is shown in Fig. 3.6(b). In Fig. 3.6(c), the rotation of the magnet is finished and the coil current is off; at this time the magnet is at rest as it was in Fig. 3.6(a). By using the swing of the Ba-ferrite magnet in this way and fixing the magnet to the parts of a wrist watch called the pallets, the watch movements can be produced.

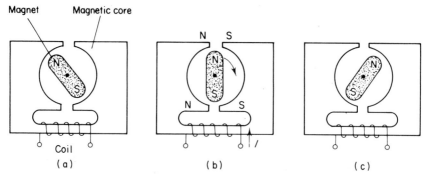

Fig. 3.6 Movement of the ferrite magnet

The pallet is controlled by a pin so that it oscillates at a fixed angle (26°) only. At the tip of the pallet a pallet jewel, which engages the escape wheel, is installed. One oscillation of the pallet sends the teeth of the escape wheel forward one circular pitch. The rotation of the escape wheel is slowed by the watch train which transmits it to the hands that indicate the time. On the other hand, there is a notch at the other end, which engages the contact jewel on the balance wheel and a reciprocal energy exchange occurs. The oscillation of the pallet causes the roller of the lever escapement to turn. In this way, the pallet works move the hands in intervals of the time standard to which the watch is set, through the function of the lever escapement.

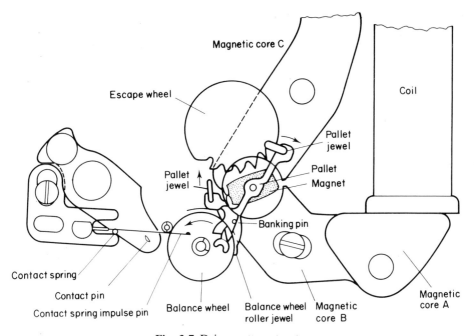

Fig. 3.7 Drive system structure

The main constituent of this Ba-ferrite magnet is $BaO \cdot 6Fe_2O_3$, and since the crystal structure is hexagonal and very long along the c-axis, it is remarkably anisotropic. Based on this, it characteristically has a very large coercive force. In what is called anisotropic Ba-ferrite, which is a parallel arrangement of crystals along this c-axis, the maximum energy $(BH)_{max}$ becomes 3.6 to 4.1 MOe, and it is widely used in places where small magnets with strong coercive force are necessary.

3.5 WHY DOES A PIEZOELECTRIC IGNITOR NOT HAVE ANY BATTERIES?

There are two types of automatic igniting devices for gas-run home appliances, battery-powered heaters and piezoelectric types. The piezoelectric type needs no batteries, has no problem of heater filament breakage and is widely used. A small version is used in cigarette lighters. This piezoelectric device uses what is called piezoelectric ceramics, and when a force is applied, the high voltage generated is discharged as an electric spark.

Fig. 3.8

This piezoelectric ceramics is composed of lead zirconate–titanate Pb(Ti–Zr)O$_3$ ceramics in which a piezoelectric property is generated through a polarization process. The ignitor device is cylindrical, with dimensions of 7 mm diameter by 15 mm length, and two of them are placed with the positive poles facing each other in the assembly as shown in Fig. 3.9. If the anvil is struck with a hammer, a high voltage of several tens of kilowatts is generated between these two poles of the ignitor. The polarity of the high voltage side is usually made positive, so a spark discharge will occur even at low voltages. When S is the surface area of the function and l is the length, and there is a striking force of F, the following equation shows the output voltage, V:

$$V = g_{33} \frac{l}{S} F$$

Here g_{33} is a voltage output coefficient related to the thickness and vibration. The units are V m/N. Also, if C is the capacitance of the piezoelectric material, the energy expended at the time of discharge is $(1/2)CV^2$. Therefore, it is necessary for C and V to be large in order to facilitate ignition. The reason for an assembly with the positive poles face to face on the high voltage side as in Fig. 3.9 is that C is increased by the parallel junction.

Also, to make V higher, it is advantageous to use materials with a large g_{33}. Also, a characteristic required of the piezoelectric material is a low fatigue development. Naturally, mechanical destruction from the hammer is not

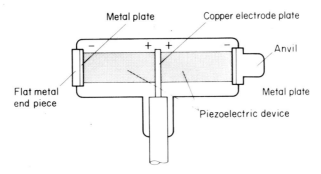

Fig. 3.9 Piezoelectric ceramic ignitor

permissible. The PZT ceramics which is currently used in ignitors is mechanically strong and rarely breaks down, even under a heavy impact. However, repeated striking causes a decrease in polarization and g_{33} becomes smaller; therefore, the voltage generated is decreased, and the material becomes useless for ignitors. On this point, however, there are reliability tests, and one hundred thousand times can be guaranteed. This one hundred thousand times means 30 times a day for 10 years. In this way, piezoelectric ignitors may be used semi-permanently without after-service.

In Table 3.2, the characteristics of the two kinds of ignitor systems are compared. In terms of smallness, igniting capacity, conservation, etc., the piezoelectric system is better.

Table 3.2 Comparison of piezoelectric igniting and battery igniting methods

	Battery method	Piezoelectric method
Ignition method	Red heat of filament	High-voltage spark between electrodes
Temperature generated	Several hundred degrees	Several thousand degrees
Voltage generated	3 V	10,000 V or more
Durability	Corrosion of filament	Semi-permanent because of heat-resistant nickel electrodes
Length of durability	6 months to 1 year	10 years or more
Ignitable gases	City gas, propane, butane	Natural gas, city gas, propane

3.6 WHAT IS A CERAMIC TRANSFORMER?

There is a piezoelectric transformer that uses a lead–titanate system material. The voltage ratio is of the order of 300 to 500. Taking into account electrical characteristics and manufacturing, the shape is generally a thin rectangular block. The structure is shown in Fig. 3.11; lengthwise, half of the lead titanate substrate forms a pair of terminals across the thickness of the block, while the end face of the other half has another terminal. The former is polarized across the thickness, and the latter is polarized lengthwise. The part polarized across the thickness is called the drive section and the other is called the generating section.

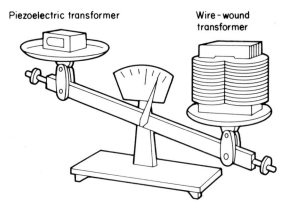

Fig. 3.10 Piezoelectric transformers are smaller and lighter

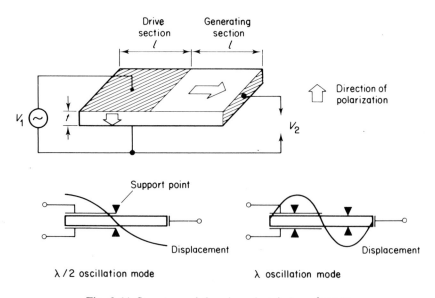

Fig. 3.11 Structure of the piezoelectric transformer

If an input voltage with a specific resonant frequency determined by the length, $2l$, of the driver section is supplied, a mechanical vibration occurs lengthwise because of electrostriction. Because of this vibration, a high-voltage output may be obtained since an electric charge is generated in the generating section because of a piezoelectric effect. This shows that the step-up process occurs through an electrical → mechanical → electrical conversion sequence. The following equation shows the step-up ratio, G_∞, when there is no load:

$$G_\infty = (V_2/V_1) = 4/\pi^2 \cdot Q_m \cdot k_{31} \cdot k_{33} \, \frac{l}{t}$$

From the above equation, the step-up ratio can be determined from the electromechanical coupling coefficient, the mechanical quality coefficient, and the shape of the sample.

The materials that meet the requirements for use in piezoelectric transformers should have large electromechanical coupling coefficients, k_{31}, k_{33}, and a mechanical quality coefficient Q_m of around 400 to 1,000. Furthermore, since they are used under conditions close to those of high amplitude vibration with a large input signal, the displacement at the point of minimum stress is extremely large, and mechanical fatigue causes a deterioration of the piezoelectric constant. Therefore, materials which have little mechanical fatigue deterioration are used for piezoelectric transformers.

The step-up function of a piezoelectric transformer works at either the frequency of the half wavelength ($\lambda/2$ mode) of the device, or that of the whole wavelength (λ mode), or in that vicinity. When the length of the device, $2l$, is 56 mm, the frequencies are 30 kHz ($\lambda/2$ mode) and 60 kHz (λ mode). The relationship between the output voltage and the frequency in this case is shown in Fig. 3.12. In general, the step-up ratio is larger for the λ mode than for the $\lambda/2$ mode, and with a load of 100 MΩ, exhibits values of 250 to 300.

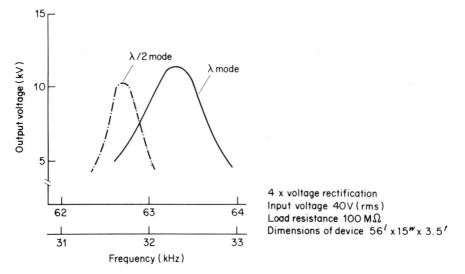

4 x voltage rectification
Input voltage 40V (rms)
Load resistance 100 MΩ
Dimensions of device 56^l x 15^w x 3.5^t

Fig. 3.12 Frequency characteristics of the piezoelectric transformer

The applications of piezoelectric transformers that are being investigated include continuous igniting devices and high-voltage generators for black and white television sets. The igniting device which uses this transformer is capable of continuous electric discharge, so ignition is 100% reliable. Because of this, they have possibilities as lighting devices for oil burners and other oil-run devices. On the other hand, use in high voltage generators for television sets must take into consideration the problem of regulation, and a method for improvement that includes putting a ZnO system ceramic varistor into the rectifier circuit has been found.

3.7 WHAT IS THE PIEZOELECTRIC RESONATOR THAT DISTINGUISHES BETWEEN THE NUMBERS ON BULLET TRAINS (*SHINKANSEN*)?

At the central surveillance center, the system-wide operating conditions on all bullet train lines can be observed. Just as marathon runners have numbers the trains are numbered too, and their locations are confirmed using an electromagnetic inductive train number identification system which is made up of an apparatus that indicates the number of the train and an apparatus that discriminates among train numbers which is set up at each station. The reason it is called an electromagnetic induction system is that the main element in the train-mounted apparatus uses a piezoelectric resonator. The electromagnetic waves used are resonant frequencies of the medium wave band, 270 to 520 kHz, and if, for example, these frequencies are made to correspond to five different numbers, 0, 1, 2, 4, and 7, an aggregation of these frequencies can express decimal numbers from 1 to 0. The identification of

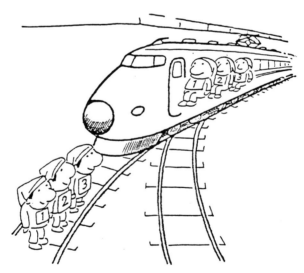

Fig. 3.13

the numbers by the frequency aggregation is accomplished by an electro-magnetic inductive coupling of station-based and train-based coils. In other words, when the sweep waves f_1 to f_5 are transmitted via the coil (transmission coil) of the station-based transmitter, electromagnetic induction in the train-based coil gives rise to the same sweep waves. When the apparatus on the train has piezoelectric resonators for just f_2 and f_3, only the instantaneous frequencies f_2 and f_3 are induced in the same way in the station-based receiving coil, detected, processed, and decoded.

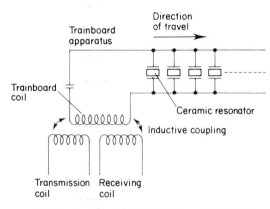

Fig. 3.14 Schematic of the number recognition and identification apparatus principles

Piezoelectric ceramic resonators used for such purposes are required to perform satisfactorily under severe conditions such as exposure to the elements, vibration, and shock. The temperature and durability characteristics needed for the resonant frequencies and directly related to number identification errors are especially severe, of the order of $\pm 0.2\%$ over a period of ten years.

Meeting these requirements called for new material development, and a $PbTiO_3–PbZrO_3–Pb(Sb,Nb)_{1/2}O_3$ system piezoelectric material with a MnO_2 additive was found. This material has a Q_m of 2,000 to 4,000, and its temperature coefficient is 5 ppm/degC or less from $-40\,°C$ to $60\,°C$.

Furthermore, in order to confirm the stability over a long period of time, it was necessary to determine the range of variation through high-temperature accelerated aging and the resulting acceleration coefficient. The results of tests of the variation rates for resonant frequency and resonant resistance on 1,000 samples at $100\,°C$ for 22,000 hours gave excellent data with an average of 0.06%. Even if an acceleration coefficient of 10 times the value is considered, confirmation of a sufficiently low rate of variation of 0.1% is possible.

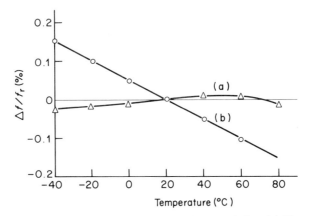

Fig. 3.15 Resonant frequency and temperature characteristics. (a) New material. (b) Products up to present

3.8 WHAT KINDS OF CERAMICS ARE USED IN THE DETECTORS OF ULTRASONOGRAPHS?

Ultrasonic devices make use of the reverberations from ultrasonic pulses. While ultrasonic pulses are propagated in the living body, the different tissues of the organs, or diseased tissues, alter the sound in varying degrees, and from these types of boundaries echoes return. During the pulse generation, the echoes are detected one by one and displayed on a CRT or some other apparatus. From this display, the distance to the target of the waves can be determined through the changes in the velocity of the transmitted waves and the time from the emission to their return. Since a directed pulse is used, the

Fig. 3.16

direction of the target can also be determined. Furthermore, by using the directed pulse with a scanning motion, the parallel pulses generate a two-dimensional ultrasonographic display. This scanning, which is carried out electronically, is called electronic scanning, and both linear and sector electronic scanning are used.

Since the material for the ultrasonic probe of this electronic scanner is the source for generating ultrasonic waves and at the same time converts ultrasonic vibrations picked up from outside into electronic signals, either a PZT or a three-component piezoelectric ceramics is used. As shown in Fig. 3.17, the ultrasonic probe is composed of many strip-like transducers with widths of 1 mm or less (piezoelectric ceramics that perform the electrical → mechanical → electrical conversion), an absorber (load material backing) to absorb the useless ultrasonic waves radiated from the back sides of the transducers, and thin coating on the surface of the transducers for their protection and for impedance matching between the body and the ultrasonic waves. The piezoelectric ceramics works in the transducer to control the characteristics of the probe and to improve the quality of the ultrasonographic image.

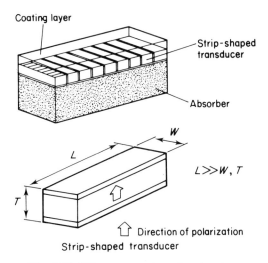

Fig. 3.17 Ultrasonograph probe structure

Since the frequency of the ultrasonic waves depends on the thickness of the ceramic substrate, and since, if the width of the strips is increased, the vibration along the width stimulates a simultaneous vibration through the thickness, the ratio of the width to the thickness must be limited to 1.0 or less to ensure an efficient vibration through the thickness. On the other hand, with progress in the use of higher frequencies from 2–3 MHz to 5 MHz, the oscillators become thin and narrow, and the difficulty of manufacture is a problem. Because of this, a 5-MHz probe with a length of 200 mm which makes use of a PbTiO$_3$ system ceramics that has a large electromechanical coupling coefficient anisotropy ($k_t \gg k_{31}$) has been developed. The induced vibration across the width is small, and even if its width to thickness ratio is large, its use is feasible.

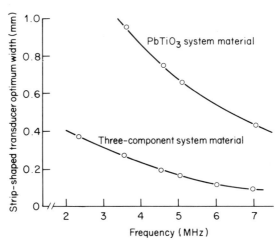

Fig. 3.18

3.9 ARE THERE CERAMICS WHICH ACT AS SUBSTITUTES FOR THERMOMETERS?

There is a semiconductor called a 'thermistor.' 'Thermistor' is an abbreviation for 'thermally sensitive resistor,' and thermistors have a large negative temperature coefficient; with a rise in temperature, the resistance of the ceramics is radically reduced. The constituents of thermistors are oxides of nickel (Ni), cobalt (Co), manganese (Mn), etc. The relationship between resistance and temperature is expressed as follows:

$$R = R_0 \exp\left(\frac{1}{T} - \frac{1}{T_0}\right) B \tag{1}$$

where R is the resistance of the thermistor at temperature T (K), R_0 the resistance at temperature T_0 (K) and B the semiconductor activation energy divided by the Boltzmann constant, generally called the thermistor constant. The larger the value of B, the larger the proportion of the change in the resistance of the thermistor to the change in temperature becomes. The materials used generally have B values of 2,000 to 6,000 K.

The temperature coefficient of a thermistor, the property it makes use of, is given by the following equation:

$$\alpha = \frac{1}{R} \cdot \frac{dR}{dT} = -\frac{B}{T^2} \tag{2}$$

In a thermistor in which $B = 3,400$ K, the value of α at 25 °C is about -0.04, which shows a negative value of about 4% per degC. This value is 10 times that of most standard metals, and one can see that the resistance versus temperature change is extremely large. Furthermore, it can be seen from equation (2) that the temperature coefficient is not uniform like that of metals, and since it is inversely proportional to the square of the absolute temperature, it is low at high temperatures and high at low temperatures. Because of this, thermistors are suitable for use at temperatures of 300 °C or less (Table 3.3).

Table 3.3 Temperature coefficients of metals and thermistors

Material	α (%/°C)	B value (K)
All-purpose thermistor	3–4 (at 27 °C)	3,000–4,000
High-temp. thermistor	2–3 (at 900 °C)	15,000–18,000
Pt wire	0.37–0.40 (at 27 °C)	
Ni wire	0.54 (at 27 °C)	
Au wire	0.34 (at 27 °C)	
Fe wire	0.50 (at 27 °C)	

Other than having a high temperature coefficient, special characteristics of thermistors include a comparatively high resistivity, and thus the magnitude of the temperature-based change in resistances is large, and a large signal may be obtained.

Fig. 3.19

The uses of thermistors are divided into the areas of temperature correction and measurement, and for the measurement type the stability is 0.2% or below during the first year and this value decreases with time to create extreme stability. Also, on the point of sensitivity, they are used in ultra-sensitive thermometers capable of measurements between 1/100 °C and 1,000 °C, and in some cases as low as 1/10,000 °C. The range of applications of thermistors is very wide; their output can be detected electrically, so they are used in temperature control and management systems.

In recent times, the control of automobile exhaust has become stricter, so temperature sensors for detecting the temperature rises associated with after-burning in the exhaust systems of vehicles that use catalytic converters and thermal reactors have become necessary. The temperatures at which the automobile sensors are used make stability at temperatures of 1,000 °C and above necessary; therefore, materials which are different from those of standard thermistors, which are limited to around 300 °C, are being developed. As shown in Table 3.3, materials which have temperature coefficients, the value of B, as large as 15,000 to 18,000 K include Y_2O_3 systems, CaO systems, CoO–Al_2O_3 systems, etc.

The conduction mechanism of these materials can be divided into two types, ionic and electronic. ZrO_2 with CaO, MgO, and Y_2O_3 additives, which is called stabilized zirconia, is most often used. The requirements for high-temperature thermistors are more severe than those for all-purpose thermistors, in that not only the temperature coefficient characteristics, but also the stability to prevent elemental changes, the absence of crystal transformations, etc., are required.

3.10 WHAT ARE THE CERAMICS THAT DETECT INFRARED LIGHT?

Lead titanate is a ceramics that can detect infrared light. The method of measurement used with this material is based on Wien's law: 'a body radiates a rather wide range of wavelengths of light in correspondence with temperature; the strongest wavelength is inversely proportional to the absolute temperature T, and the total radiation is proportional to T^4.' In effect, the method for detecting the amount of infrared radiation uses a slight rise in temperature caused by a sensing of the energy associated with radiated infrared light.

Lead titanate ceramics is a constituent of the piezoelectric ceramics called PZT and is a perovskite-structured ferroelectric substance of the same type as barium titanate, but there has been no opportunity to use it by itself. This is because the large anisotropy of lead titanate crystals causes trouble, making sintering very difficult and high-quality ceramics hard to obtain. However, in recent years sintering technology has advanced, and a search for additives has been carried out, so that a satisfactory durable ceramics can be obtained. Since lead titanate is a ferroelectric substance, spontaneous polarization occurs below the Curie temperature of 490 °C. However, because this spontaneous polarization is dependent on temperature, the surface charge generated by it changes accordingly; therefore, changes in temperature can be detected by changes in voltage. The detecting device is made from 20–30 μm thick polarized lead titanate ceramics, with terminals made by vaporizing NiCr or AuCr. Leads are taken off of the terminals; an input resistance is created by an FET (field effect transistor) circuit impedance transformation, and it is amplified.

Ceramic sensor

Fig. 3.20

The good things about this detector are that, unlike a thermistor, it does not make use of resistance changes, and since it detects by means of voltage changes, a bias voltage, which could be a source of noise, is not necessary. The voltage or electric current due to this kind of pyroelectric effect is extremely low and the resistance of the device must be 10^{12} Ω or greater for effective detection of infrared radiation. Furthermore, the device works as a capacity-measuring device, even with its low-frequency voltage which is close to being a direct current. If it is operated with a sufficiently low load, high-speed detection and excellent pulse characteristics may be easily obtained.

Using wavelengths starting with several micrometers, in other words the far infrared spectrum, applications extend from ocean and land measurements by satellites and aircraft to industrial applications such as power-line transformers and infrared burglar alarms, to medical applications such as no-touch skin measurements, to measurement of pollutant gases such as CO and SO_2, etc.

In Table 3.4, the properties of several pyroelectric materials are shown. It can be seen that the properties of lead titanate ceramics are excellent.

Table 3.4 Characteristics of pyroelectric materials

Pyroelectric material	Curie temp. $T_c(°C)$	dP_s/dT ($\mu C/cm^2 K$)	Permittivity ε	Volume specific heat c ($J/cm^2 K$)	Absorbed wavelength range λ (μm)
TGS (crystal)	49	0.04	35	2.5	0.25–2
$Sr_{0.5}Ba_{0.5}Nb_2O_6$ (crystal)	120	0.065	380	2.1	0.5–6
PZT-5A (ceramics)	365	0.04	1,900	3.1	
$PbTiO_3$ (ceramics)	470	0.06	200	3.2	0.6–6
$LiTiO_3$ (crystal)	660	0.023	54	3.15	0.5–6

Pyroelectric lead titanate ceramics with small amounts of MnO and La_2O_3 added is a material with an excellent Curie temperature, dielectric constant, dielectric loss, pyroelectric coefficient, etc., along with excellent machinability and long-range after-effects. Good machinability is especially necessary in the manufacture of pyroelectric detectors, since the fabrication processes include the making of a lead titanate ceramics wafer with a thickness of around 20–30 μm and a polarization process involving the application of a 100 kV/cm electric field at 200 °C. Also, depending on the mixture of the additives, this ceramics can easily have high resistance and low thermal conductivity, and furthermore there are many other excellent aspects such as no cleavage in single crystals, etc. When it is used as a temperature sensor, with an optimized measuring circuit, it is possible to measure temperature fluctuations to 10^{-5} °C, and the development of ferroelectric materials with Curie points in the normal temperature range will make wide applications possible.

3.11 WHAT KINDS OF CERAMIC HEATING ELEMENTS ARE THERE?

There are silicon carbide, molybdenum silicide, lanthanum chromite and zirconia ceramic heaters. Since silicon carbonide decomposes and sublimes at standard pressure, it has no softening point. Its decomposition temperature is 2,400 °C. Molybdenum silicide, the chemical formula of which is $MoSi_2$, has a very good resistance to oxidization, and its melting point is 2,030 °C. On the other hand, the melting point of zirconia is 2,690 °C and that of lanthanum chromite is 2,490 °C. In this way, most of the ceramics used for heat generation have high melting points, and this is because the higher the melting point, the higher the maximum usable temperature. Zirconia, especially, can be used safely up to a temperature of 2,000 °C.

However, a high melting point is not the only prerequisite for a material to be used as a heating element; the conditions also include its having an appropriate resistivity for converting electricity into heat. If a voltage is applied to a substance with a certain resistance, it is commonly known that energy proportional to the square of the current running through the substance will be converted into heat which is called Joule heat. This is the functional heat of the heating element; but in order to generate it, the resistivity must be several Ω cm or less. On the other hand, the resistivity peculiar to a certain substance has certain temperature characteristics, and these characterize the way it is used for heat generation. Since SiC has positive temperature characteristics from around 600 °C and above, it is characterized by easy temperature control. Molybdenum silicide has positive temperature characteristics from room temperature up, and its temperature coefficient is small at 0.0048. The resistivity of lanthanum chromite is 0.14 Ω cm at 500 °C, but it decreases to 0.11 Ω cm at 1,800 °C. The case of zirconia is especially remarkable. Its resistivity is an extremely high 10^7 Ω cm at 500 °C, but decreases to 180 Ω cm above 1,000 °C, 1.9 Ω cm at 1,500 °C and 0.4 Ω cm at 2,000 °C. Although most of these heating elements become rather good conductors at 1,000 °C and above, zirconia can be used at ultra-high temperatures where it is impossible to use other heating elements, because it maintains a low resistance above 2,000 °C. However, there is an inconvenient aspect. Since it has a high resistance below 1,000 °C, it doesn't function as a heater, and preheating up to the temperature at which it shows conductivity by another heating element is necessary.

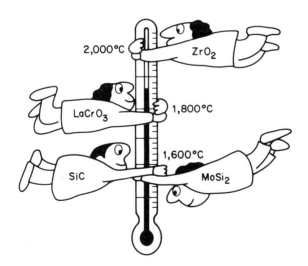

Fig. 3.21 Maximum temperatures for stable use of various heating elements

In order to improve the anti-oxidation characteristics, corrosion-resistance and mechanical strength of these ceramic heating elements, additives and partial elemental substitutions have been tried. For example, in the vicinity of 1,190 °C zirconia undergoes a reversible monoclinic $\rightleftarrows$ tetragonal transformation, and since a collapsing phenomenon is caused, studies of stable and semi-stable phases made with CaO, MgO, Y_2O_3, etc., additives have been carried out. In lanthanum chromite, a partial substitution of Ca for La lowers the resistivity, and an electric current will pass through from room temperature on up.

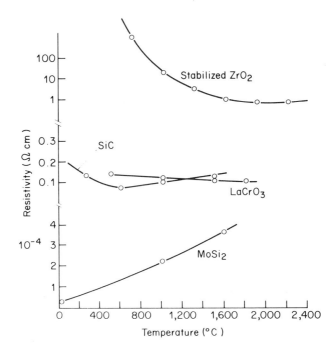

Fig. 3.22 Temperature vs. electrical resistance characteristics of various heating elements

3.12 WHAT ELECTRIC CHARGE CAN BE STORED IN CERAMICS?

If a voltage is applied across two parallel electrodes in air, it can be assumed that there will be no polarized charge in the air, and the following equation gives the capacitance:

$$C_0 = 8.854 \times 10^{12} \frac{A}{d} \text{ (F/m)}$$

Here A (m^2) is the area of the plates and d (m) is the space between them.

Next, if the space is replaced by a material with permittivity ε, the capacitance increases to εC_0. From this it follows that capacitance is proportional to the area of the plates and the permittivity, but is inversely proportional to the separation of the plates. Therefore, to the extent that the permittivity and the surface area of the dielectric material get larger, and the material is made thin, the larger the capacitance becomes.

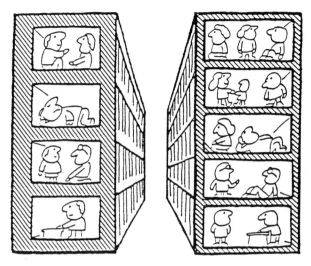

Fig. 3.23 In two condominiums of the same size the number (volume) of people who can live there is less in the one with thicker walls

On the other hand, the microstructure of ceramics is composed of crystallites and grain boundaries. A BL condenser is a large capacitance condenser that makes positive use of this characteristic of ceramics. BL is an abbreviation for 'boundary layer.' The structure of this condenser is shown in Fig. 3.24(a) and an equivalent circuit schematic is shown in Fig. 3.24(b). The crystallites are semiconductive and their resistivity is 10^0 to 10^3 Ω cm, while the grain boundaries are insulating and form an extremely thin dielectric layer about 0.3–2 μm thick.

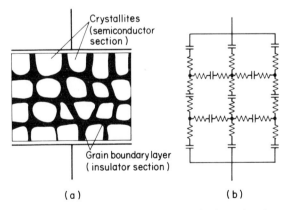

Fig. 3.24 Microstructure and its equivalent circuit

The basic materials of a BL condenser are barium titanate (BaTiO$_3$) and strontium titanate (SrTiO$_3$). These substances are called 'ferroelectrics,' and their permittivity is temperature-dependent and is at a maximum at which is called the Curie point, the phase transition point. Addition of elements with different valences to this substance generates free electrons which are easily moved in an electric field. Furthermore, if it is sintered in a reducing atmosphere, oxygen deficiencies are generated and the resistance is lowered. Metal oxides such as MnO_2, Bi_2O_3, CuO, Ti_2O_3, Sb_2O_3, etc., are applied to the surface of semiconductive ceramics obtained in this way, and the desired insulation is created in the spaces between the crystallites by thermal diffusion. If this grain boundary layer is made thinner, an increase in capacitance is obtained; therefore, grain growth is facilitated and crystallites 100–120 μm in diameter are obtained. In the $BaTiO_3$ system obtained in this way, the capacitance is 0.1–0.5 μF/cm^2, tan δ 5%, and IR 500 MΩ/cm^2. For the $SrTiO_3$ system, the values are 0.08 μF/cm^2, tan δ 1% or less, and IR 500–1,000 MΩ/cm^2.

These BL condensers are excellent condensers that combine an approximately 1 GHz central dispersion frequency with high permittivity. These condensers perform with notable efficiency as wide-band bypass capacitors in the emitter circuits of 12 MHz FDM super-multiplex transmission relays, and with several types of improvements are used in various ways as wide-band bypass capacitors up to the several gigahertz range.

With its outstanding characteristics this BL condenser not only contributed to the improvement of devices by its use, but also imparted a great influence to the development of ceramic electronic parts. This is the positive use of the microstructure of ceramics (crystallites and grain boundaries), which is the exact opposite of the thinking that led to attempts to get close to single-crystal structures that preceded it. The ZnO varistor discussed in Section 3.1 makes use of this idea.

3.13 WHAT ARE ELECTRO-OPTICAL CERAMICS? HOW ARE THEY USED?

The researchers at Sandia Laboratories in the USA had been trying for some time to hot press lead zircon-titanate, $Pb(Zr,Ti)O_3$, a ceramics that had been well-known for a while, in which a partial substitution for Pb had been carried out. Around 1968, Haertling and coworkers found that even though the Bi substitution form of this ceramics was polycrystalline, it showed high optical conductivity, and they also found that the optical characteristics were dependent on the direction of polarization. While they continued to raise the optical conductivity, they found similar results with other substitutions. Furthermore, in 1971, they found that $(Pb,La)(Zr,Ti)O_3$, which is called PLZT ceramics and is $Pb(Zr,Ti)O_3$ with a La_2O_3 additive, showed extremely low light-scattering, high optical conductivity, and optical anisotropy in an electric field. In this way, translucent piezoelectric ceramics, which had until then been limited to single crystals, took a place among electro-optical materials, and now has great potential as a material for devices with electro-optical functions.

When the Pb^{2+} ions of PLZT translucent piezoelectric ceramics, $Pb(Zr,Ti)O_3$, are partially replaced by La^{3+} ions the chemical formula changes according to the lattice vacancies that are produced in the perovskite-type structure. Although not completely confirmed at present, the following is normally used:

$$[Pb_{1-x}La_x][(Zr_yTi_z)_{1-x/4}]O_3 \ (y + z = 1)$$

$PLZT_{x/y/z}$ is commonly written to express the PLZT composition.

In order to improve the optical conductivity of ceramics, chemical uniformity and reduction of residual pores are necessary. Up to now, the methods for manufacturing PLZT that have been reported are as follows: (i) hot pressing, (ii) sintering in a PbO atmosphere, (iii) hot isostatic pressing, and (iv) multi-stage sintering. Among these methods, hot pressing in an apparatus like the one shown in Chapter 1, Fig. 1.18, is generally used. In general, it is obtained by hot pressing for several hours at 1,000–1,300 °C under a pressure of 200 to 600 kg/cm².

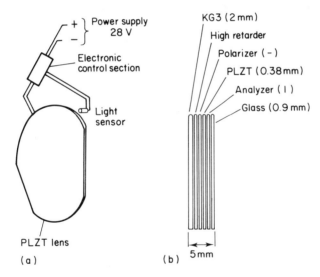

Fig. 3.25 Principle and structure of antiglare lenses. (a) Fundamental schematic. (b) Structure

There are many kinds of proposals for applications of translucent piezoelectric ceramics in electro-optical devices, almost all of which make use of PLZT. PLZT picture storage and display apparatus include both double-refraction mode and light-scattering mode usages. Of the former, Ferpic and bias-striction-type Ferpic were proposed by the researchers at Bell Laboratories. Of the latter, Cerampic was proposed by Sandia Laboratories. These are shown in Fig. 3.26. Furthermore, optical memory, display, optical shutters, etc., are fields in which PLZT has applications. Examples of applications of optical shutters include shutters for scanning electron microscope stereocameras, optical modulators, medical stereo-television, many kinds of industrial devices, light controls in welding monitors, and stereo observation of the ocean floor for marine development. A new application of PLZT is the transparent piezoelectric speaker. It has transparent electrodes covering both surfaces of a thin sheet of transparent piezoelectric ceramics, and it can be used for small portable apparatus such as electronic calculators and watches because of its transparency.

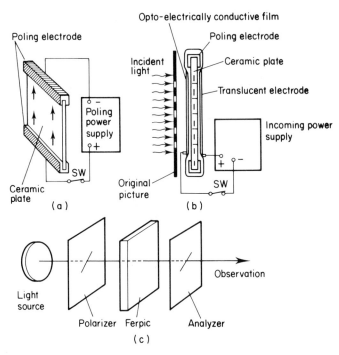

Fig. 3.26 Principles and structure of Ferpic and its display method

3.14 WHAT KINDS OF CERAMICS CAN DETECT GAS?

In general for the gas-sensors used for detecting gas leaks, direct electric outputs are suitable, and they are classified into two types, the semiconductor and the contact-burning methods. In the semiconductor method, a high temperature is maintained in a semiconductor (usually a n-type oxide semiconductor) by a heater, and when it comes into contact with a combustible gas, the change (decrease) in the electrical resistance is made use of. On the other hand, the contact-burning method makes use of the rise in temperature in a platinum wire when a combustible gas is burned by means of a catalyst in contact with it, and a change (rise) in the resistance of the platinum wire allows detection. Special characteristics include the need for a large gas concentration with the semiconductor type, and the proportionality of the output of the contact-burning type to the gas concentration. Whatever the method, the major conditions for gas-sensors are: (i) high sensitivity to the gas in question, (ii) lack of response to gases other than the combustible gas, and (iii) long-term stability of characteristics.

In developing gas-sensors, the requirements mentioned above must always be taken into account, and recently the semiconductor gas-sensor has undergone a great deal of improvement. The sensor which is currently used for combustible gas and is commercially available is made of a mixture of tin oxide (SnO_2) and minute amounts of precious metals such as palladium (Pd). This sensor shows a large response to several kinds of gases at comparatively low temperatures ($< 350\,°C$), and the SnO_2 system gas-sensor is still being studied actively, even though it has the weak point of being highly sensitive to alcohol and moisture.

As shown in Table 3.5, many kinds of semiconductor gas-sensors besides the SnO_2 system have been proposed, and among them the ZnO system has stirred up interest because of its gas sensitivity. The ZnO system gas-sensor is also an indirect heating type and has the internal heater structure shown in Fig. 3.27. Other than this kind, there are direct heating-type sensors with internal heater structures.

Table 3.5

Detector	Material detected
ZnO thin film	Reductive, oxidizing gases
Thin film oxides (ZnO, SnO_2, Fe_2O_3, NiO, etc.)	Reductive, oxidizing gases
SnO_2	Combustible gases
In_2O_3 + Pt	H_2, hydrogen carbide
Oxides (WO_3, MoO_3, Cr_2O_3, etc.) + catalysts (Pt, Ir, Rh, Pd, etc.)	Reductive gases
SnO_2 + Pd	Reductive gases
SnO_2 + Sb_2O_3 + Au	Reductive gases
WO_3 + Pt	H_2
Complex oxides ($LaNiO_3$, etc.)	C_2H_5OH, etc.
V_2O_5 + Ag	NO_2
CoO	O_2
ZnO + Pt	C_3H_8, C_4H_{10}, etc.
ZnO + Pd	H_2, CO
$MgFe_2O_4$	Reductive gases
γ-Fe_2O_3	C_3H_8, C_4H_{10}, etc.
SnO_2 + ThO_2	CO

In the ZnO system gas sensor, gas selectivity is obtained through the selection of catalysts. The sensor which uses catalysts containing Pt compounds shows a strong sensitivity to gases such as CO, H_2, etc.

As an improvement over the sensitivity of standard semiconductor gas-sensors, there are γ-Fe_2O_3 systems which, in contrast to the Pt, Pd catalytic types, are sensitive enough without the use of catalysts. Because of the absence of catalysts, which become useless, it is said to have a usage life of 50,000 or more hours.

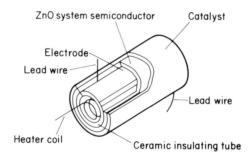

Fig. 3.27 Structure of a ZnO system gas-sensor

Even though semiconductor gas-sensors are used as gas leak alarms and are commercially available, the technology cannot be said to be complete. The reasons are that, since all of them make use of surface phenomena, the manufacturing processes and many other factors involving usage have complex effects, and uniformly stable operation is difficult to obtain. Furthermore, the fundamental research was slow in giving results, and development went ahead on a trial and error basis. Since gas-sensors have a realtion to people's lives, there is a pressing need for technical results. It is necessary for public and private research to come together on it.

3.15   WHAT KINDS OF MOISTURE-DETECTING CERAMICS ARE THERE?

As shown in Table 3.6, the history of the development of moisture sensors has extended to electrolytes of which lithium chloride is representative, the use of the hydrophile property and swelling with high-polymer organic materials, and the use of absorption and deabsorption phenomena with metallic semiconductors (Se, Ge, etc.) and metallic oxides. Since the reporting of the magnetite (Fe_3O_4) colloid device in 1966, there has been research into various materials for the moisture sensors that use ceramics, from the use of chromic oxide (Cr_2O_3), ferric oxide (Fe_2O_3), alumina (Al_2O_3), zinc oxide, etc., painted films, to the recent Fe_2O_3-K_2O system ceramics, ZnO-Li_2O-V_2O_5 system ceramics, Cr_2O_3 glazed thick film, and $MgCr_2O_4$ system ceramics, but the outputs of all of them correspond to the relative humidity, and mainly measure by a change in resistance that accompanies absorption and deabsorption.

Because ceramic moisture sensors make measurements by means of changes in resistance, the measurement circuits are simple, and most of the devices are low in cost and usable as home appliances. Most of the effort that

Table 3.6 Various moisture-sensors

Classification	Moisture-sensitive material	Detection method	Detection temp.	Period of use
Electrolysis systems	LiCl + polyvinyl alcohol	Resistance	%RH	1938–
	Polystyrene sulfide film	Resistance	%RH	1955–
	Phosphorus pentoxide film	Resistance	%RH	1955
	Potassium metaphosphoric acid	Quantity of electricity	ppm	1958–
	Saturated solution of LiCl	Resistance (temp.)	dew point	1954
Organic compound systems	Cellulose + carbon	Resistance	%RH	1954
	Cellulose butyl acetate	Volume	%RH	1969–
	Resin + carbon	Resistance	%RH (switching)	1974–
	Polyamide + quartz oscillator	Resonant frequency	%RH	1975
	Vaporized Se film	Resistance	%RH	1957
	Vaporized Ce film	Resistance	%RH	1969
	Sintered Si film	Resistance	%RH	1971–
	Al_2O_3 film	Volume and impedance	%RH	1955–
Metal and metal oxide systems	Painted Fe_3O_4 colloid film			1966
	Cr_2O_3, Ni_2O_3, Fe_2O_3, Al_2O_3, ZnO painted film ($Ni_{1-x}Fe_{2+x}O_4$) ceramics	Resistance	%RH	1974–
	Glass–ceramics thick film	Resistance	%RH	1975–
	Fe_2O_3–K_2O ceramics	Volume	%RH	1975
	ZnO–Li_2O–V_2O_5 ceramics	Resistance	%RH	1976
	Cr_2O_3 system glazed thick film	Resistance	%RH (switching)	
	$MgCr_2O_4$ system ceramics	Resistance	%RH	1976–

has gone into the development of moisture sensors has centered on ceramics, but here, the operational principles and practical applications will be looked at in terms of an example, the recently developed ZnO system.

As shown in Fig. 3.28, the structure of the sensor is a wafer-shaped high porosity ceramics both sides of which have high-porosity electrodes baked on. To these, platinum–rhodium wires are welded, and the whole is attached to a hermetic seal support. Leads are welded on, and it is installed in a rectangular plastic case and fixed with resin. It is a construction that can sufficiently endure vibration and drop tests.

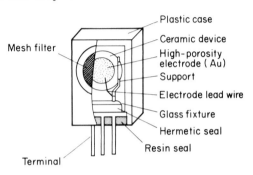

Fig. 3.28 Structure of the ZnO–Cr_2O_3 system moisture-sensor

The principles behind the moisture sensitivity and mechanism may be thought of as follows. Fig. 3.29 shows a cross-section of the structure of the ceramic component. The base ceramics consists of high porosity, spinel-structured grains made from 2–3 μm $ZnCr_2O_4$ particles, and the surface of these particles is coated with a uniform thin film glassy layer made from $LiZnVO_4$. The Li–O points, which are sensitive to humidity, are fixed in the V–O matrix structure, so the most sensitive layer has a stable structure. The surface conditions of the moisture-sensitive layer are that it holds stable OH radicals, and over these radicals a multi-molecular water molecule absorption layer is formed. It can be thought of as showing conductivity of moisture.

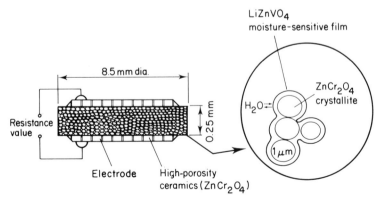

Fig. 3.29 Conceptualization of the ZnO–Cr_2O_4 system moisture-sensor

3.16 WHAT ARE THE CERAMICS THAT DETECT PRESSURE?

The effects of pressure on human actions is great, and there is a large potential need for pressure-sensors in things like measurements of atmospheric pressure in meteorological phenomena, medical measurements of blood pressure and cranial muscles, oil pressure in factories, measure and control of gas pressure, etc. The development of pressure-sensors has been slow compared with that of temperature-sensors, and their use has not gone beyond measurement.

As shown in Table 3.7, there are semiconductor, piezoelectric, and ferromagnetic solid devices for detecting pressure, and, in terms of ceramics, piezoelectric ceramics and ferromagnetic ferrites can be cited. Here, piezoelectric pressure- and acceleration-sensors, which are ahead of the others in terms of practical use, will be mentioned.

Table 3.7 Various pressure-sensors

Device	Detection method	Structure and materials
Semiconductors	Piezoelectric resistance effects	Lamination type (Si, Ge)
		Vaporization type
		Diffusion type: diffusion leads vaporization leads
	p–n junction electron transfer changes	Diode
Dielectrics	Permittivity changes Piezoelectric effects Pyroelectric effects	PZT, $BaTiO_3$ $Pb(Ti,Zr)O_3$ PUF_3 electret
Magnetic substances	Permittivity changes	Ferroelectric ferrite

(1) Piezoelectric pressure-sensors

When a force is applied to crystals such as quartz, barium titanate ($BaTiO_3$), or zinc oxide (ZnO), an output signal is caused by a polarization that depends on the pressure change, and piezoelectric pressure-sensors make use of this. They have been in use for a while, but the materials were $BaTiO_3$ and PZT ($PbTiO_3$–$PbZrO_3$) systems. Recently, however, there has been developed a piezoelectric sensor in which ZnO is applied to a substrate by a vaporization method in use, and there is a flexible piezoelectric material based on high-polymer materials such as polyvinylidene fluoride (PVF_2).

Also recently, SAW-type pressure-sensors that make use of surface elastic waves have been valued for their precision and practicality and are undergoing rapid development. As is shown in Fig. 3.30, the structure is made up of two sections, each of which has two sets of paired delay lines on a piezoelectric substrate. The sensing section has a pressure sensitive diaphragm 5–6 mm thick, and the thickness is changed according to the rating. It is made with one of the two set delay line sections at the center of the diaphragm and the other at its edge. Pressure is detected by a synthesis of the outputs of the oscillators made up of these two sets of SAW delay lines. An example of this type of sensor might have a change in frequency of 50 kHz with a pressure range of 0 to 600 psi (1 psi = 0.07 kg/cm^2)—that is 83 Hz per psi.

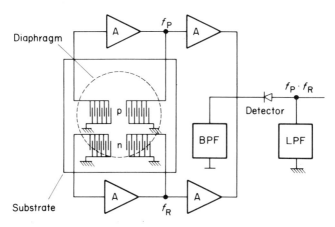

Fig. 3.30 Schematic of the SAW-type pressure-sensor

(2) Acceleration-sensor

The sensor mentioned above is for detecting static pressure, but there is a sensor for detecting dynamic force, in other words acceleration. The device used in the acceleration sensor is centered around piezoelectric ceramics. In the piezoelectric device, a number of relationships between applied forces and generated charges or voltages in the direction of application can be seen. In effect, if a striction is caused in a polarized piezoelectric ceramics through the application of an outside force, an electric charge proportional to that force is generated on the surface of the ceramics, and by means of the voltage, an observation can be made.

As noted above, the acceleration-sensor detects pressure by means of an electric charge or voltage, and its characteristics include extremely high voltage sensitivity. It has the weak point of not having static sensitivity, since an electric charge cannot be maintained because of the external limits of the charge–response type. On the other hand, since there is no trouble with hysteresis, and since amplification is easy because of no datum fluctuation, it is very easy to use.

3.17 WHAT ARE THE USES OF TRANSLUCENT ALUMINA?

In general ceramics have been considered unfit for optical materials since they are not transparent, but currently, as was mentioned in Section 3.13, many kinds of translucent ceramics are being developed, especially translucent alumina, which was the first translucent ceramics, developed by GE researchers in 1958. Since it is superior to glass in thermal resistance and strength, it has been used in the lumination tubes of high-intensity discharge lamps and high-temperature translucent receptacles. There has been especially remarkable progress in the lumination tubes for HID (high-intensity discharge) lamps. Therefore, improvements have been made from the points of view of efficiency and the lengthening of the life of HID lamps. Since the life of the HID lamp depends more on the sealing solder than on the alumina used for the lumination tube, the requirements center on how to raise the transmissivity. Fig. 3.31 shows a high-intensity sodium lamp and lumination tubes.

Fig. 3.31 High-intensity sodium lamp and lumination tubes

The following points become problems where the passage of light through ceramics is concerned. First, the absorption of light caused by impurities contributes the most to the lowering of the transmission coefficient. Next, in the same ways as were mentioned in Section 3.13, there is reflection of light and scattering caused by surfaces, grain boundaries and pores, and this reflection and scattering lengthens the light paths, with a resulting increase in the amount of light absorption. Correspondingly, the improvement of transparent alumina begins with making it highly pure and dense.

With high purification and high densification, light-absorption and light-scattering by pores is decreased, and there is an increase in the diffuse transmissivity. The diffuse transmissivity is defined as the ratio of the total amount of light emerging from the lumination tube, with its light-scattering by surfaces, grain boundaries, and pores, to the total amount of light inside the tube. Materials with a ratio of about 90% or greater can be used for HID lamps, and recently, 97% materials have been obtained.

However, the light that is scattered and reflected inside or by the surface of the alumina ceramics and goes back inside the tube, is absorbed by the sodium vapor, and as noted above, the result is an influence on the diffuse transmissivity. Recently therefore, improvement of the linear transmissivity is required. The linear transmissivity correlates with the transparency that can be seen with the eye, and since it is the ratio of emergent light unaffected by reflection and scattering to the generated light, pores, grain boundary, and surface conditions become problematic.

If the amount of pores is decreased and the surface is polished enough, it can be seen that grain-size exerts an influence on the linear transmissivity. Increasing the grain-size is effective in decreasing the light-scattering caused by grain boundaries, but on the other hand, delicate control of grain-size is necessary, because the strength of the ceramics is inversely correlated with the grain-size. An example of the relationships between microstructure and various characteristics is shown in Fig. 3.32 and Table 3.8. From these data, it can be seen that the surface conditions created by a composition with both large and small grains, as is shown in Fig. 3.32(c), are not satisfactory. Material with an average grain-size of 20 μm and a size uniformity of 1.8 or less are suited for use in lumination tubes.

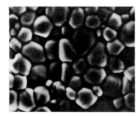 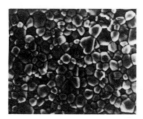 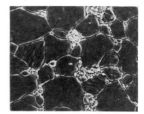

Fig. 3.32 Microstructure of translucent alumina cermaics. (a) Uniform large grain structure. (b) Uniform small grain structure. (c) Structure with a variety of grain sizes

Table 3.8 Properties and microstructure of alumina ceramics

Property	Sample (a)	(b)	(c)
Relative density	3.98	3.98	3.98
Average grain size D_a (μm)	26	13.5	28
Largest grain size D_m (μm)	44	21	120
Uniformity D_m/D_a	1.7	1.6	4.3
Pressure tube strength (N/mm^2)	250	450	135
Overall transmissivity (%)	96.5	95	93.5
Linear transmissivity (%)	10	5	6

4 Glass and Optical Fibers: Questions and Answers

Chapter 4 is concerned with glass and optical fibers. First, the differences between crystals and glass, what crystallized glass is, and the basic facts about opal glass, silica glass, Vycor glass, and various other types of glass will be touched upon. In terms of the workings of glass and radiation, glass dosimeters become a topic for discussion. Also, another big topic recently is the glass used in the processing of radioactive waste. Furthermore, laser glass, the colored glass used in photographic filters, etc. are discussed. And since just about anything said about this field is an important topic, the glass fibers used in optical telecommunications will be discussed under optical fibers.

4.1 GLASS IS SAID TO BE NON-CRYSTALLINE; WHAT IS THE DIFFERENCE BETWEEN CRYSTALS AND GLASS?

It may be said that the structure of glass is between that of a liquid and that of a solid. In the past, the structure of glass was considered to be like that of a frozen liquid, but now, a structural theory closer to that of crystals has been developed. If the structure is defined in terms of the manufacturing process, glass is in the condition of a liquid that has been melted at high temperatures and then solidified in a cooling process in which there is no crystallization. During the cooling of glass, the point that clearly distinguishes it from liquids is the existence of a transition region, and practically speaking, solids at temperatures above the transition point at which the heated liquid solidifies (liquids with high viscosity) can be distinguished from solids cooled down through the transition point. The former are supercooled liquids, and the latter are glass in a strict sense. That is, it is characteristic of glass that it always has a transition point.

A conceptual pattern of the glass structure is shown in Fig. 4.1(a). The SiO_4 tetrahedra that make up the basic structure pretty much resemble crystals, but between any two tetrahedra there can be a joint vertex, a bridge built by oxygen (oxygen bridge). As is shown in Fig. 4.1(b), it is clear from the results of X-ray diffraction analysis that angles of the joint vertices are distributed between 120° and 180°.

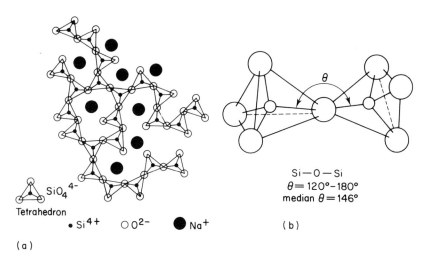

Si—O—Si
$\theta = 120°-180°$
median $\theta = 146°$

(b)

Tetrahedron

$SiO_4{}^{4-}$

• Si^{4+} ○ O^{2-} ● Na^+

(a)

Fig. 4.1 Structural models of glass. (a) Soda-silica glass. (b) Bonding in the SiO_4 tetrahedron

In silica glass, it is known that the median of the distribution of this angle is 146°. When Na_2O is introduced into the SiO_2 system (soda-silica glass), an

$$-\overset{|}{\underset{|}{Si}}-O-\overset{|}{\underset{|}{Si}}- \; + \; Na-O-Na \rightarrow (-\overset{|}{\underset{|}{Si}}-O^- \ldots\ldots\ldots Na^+)_2$$

reaction takes place, and some of the oxygen bridges are broken forming an $Na^+ \ldots O^-$—Si— arrangement of opposite charges. As can be seen in Fig. 4.1(a) this glass takes the (modified) form of net meshes made up of 5 to 8 SiO_4 tetrahedra, with Na^+ in the gaps.

Zacharriasen's rule for glass include the following four items: (i) each oxygen ion should not be linked to more than two cations; (ii) the coordination number of the oxygen ions about the cation must be four or less; (iii) oxygen polyhedra can share vertices with each other but cannot share edges or faces; (iv) at least three vertices of each polyhedron should be shared with another.

However, these rules are too strict, and as with phosphate glass, examples with two shared vertices, as well as cases like non-oxide glass where these rules are not applicable, can be found. At any rate, the fundamental structures of glass are being re-examined because of the development of the technology of ultra-rapid cooling.

Study of X-ray diffraction structure analysis has continued since some time ago, but a direct explanation of the glass structure has not been established. However, evidence for a substance being a glass is often obtained by confirming these plots, and some examples are shown in Fig. 4.2. They are the X-ray diffraction patterns of glass composed of SiO_2, silica gel, and cristobalite (crystal), and they show that a typical glass shows a pattern of one centrally located sharp rise that can be seen in silica glass. A similar pattern can also be seen in amorphous metals or glassy metals.

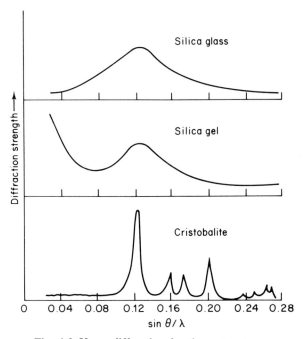

Fig. 4.2 X-ray diffraction for three structures

Besides the fusion method mentioned above, the following fine glass manufacturing methods are current topics. Hydrolysis of metal alkoxide complexes, vacuum evaporation, cathode sputtering, chemical vapor deposition (CVD), glow discharge, and hydrolysis all form glass at temperatures lower than the melting temperature.

4.2 WHAT IS OPAL GLASS?

Opal glass is the name for a man-made emulsification that resembles natural opal, and its structure is one in which particles with a refractive indices different from that of the glass are diffused in the glass. The gemstone which is called opal is composed of a hydrous colloid of silica, $SiO_2 \cdot nH_2O$ which is amorphous. The naturally formed spherulites form themselves into a diffraction grating and diffract light beautifully. The structure of opal glass is not so highly controlled as to form a diffraction grating, so it does not perform a high-level function.

Opal glass is made from transparent liquid glass, in which, through a cooling process, liquid drops or particles with a different refractive index are separated out, and because of their existence, light is scattered. The light-scattering capacity is expressed in terms of the wavelength of the incident light, the radius of the particles and the ratio of the refractive indices of the particles and the glass, $m = n(\text{particle})/n(\text{glass})$. The maximum scattering is given when the diameter of the particle $d_{max} = 4.1\lambda/2\pi\ (m - 1)$. The scattering coefficient S is expressed by $S = (3/4)\ KV_p/r$. Here, V_p is the volume fraction of the particles and K is a scattering factor. The scattering coefficient S is also called the turbidity coefficient, and if I_0 is the incident light and I the exiting light, $I/I_0 = \exp(-Sx)$ (x is the thickness through which the light must travel).

The types of compounds used to form these types of particles are shown in Table 4.1. Particles with a slow dissolution velocity are used as additives in the glass, and by remaining liquid they form a suspension. An example of the way this glass is made is tin oxide opal glass in which, in these proportions, SiO_2 37.7, PbO 28.9, K_2O 10.0, SnO_2 23.4 and a small amount of phosphate would often be added. Melting takes place at 1,300–1,400 °C, where a uniform liquid from which the glass can be formed is obtained. On cooling, it becomes slightly milky, but if it is thermally treated by reheating and maintaining it at 700–800 °C, a white, milky opal glass may be obtained. When this glass is used for craftwork, the thermal treatment is varied locally so that transparent sections and a gradation of milky parts can be made distinguishable from each other. In general, the higher the treatment temperature is, the larger and fewer the precipitated particles become. It changes with respect to the constituents, but with rapid cooling from 1,100 °C, a transparent glass-like conditioned is obtained. Since many particles of the order of the wavelengths of light are precipitated out by thermal treatment at 700 °C, this is an appropriate range.

Opal characteristics are exhibited not only when the particles are frost-like crystals, but also when they are glassy drops (in other words, phase-separation) formed in the glass. In the real substance, either the phase-separation areas themselves or the crystallites formed by means of the phase-separation are mixed in.

When the particle size is larger than the wavelengths of light, the material is called alabaster glass and is distinguished from opal glass. In addition, dispersion particles used to form opal glass are arranged in Table 4.1. Here, the ratio of refractive indices is shown assuming a refractive index of $n = 1.50$ for the glass, and it is related to the light-scattering function as mentioned above.

Table 4.1 Classifications of opal glass

Types of diffraction particles	Composition of precipitates	Diffraction coefficient (glass = 1.50)	Ratio of diffraction coefficients n-particle/n-glass
Additives	SnO_2	2.0	1.33
	$ZrSiO_4$	2.0	1.33
	ZrO_2	2.2	1.47
	ZnS	2.4	1.6
	TiO_2	2.52–2.76	1.68–1.84
Precipitated crystals	NaF	1.3	0.87
	CaF_2	1.4	0.93
	$CaTiSiO_5$	1.9	1.27
	ZrO_2	2.2	1.27
	$CaTiO_3$	2:35	1.57
	TiO_2	2.52–2.76	1.68–1.84
Inert precipitates	Air bubbles	1.0	0.67
	As_2O_5	2.2	1.47
	$PbAs_2O_6$	2.2	1.47
	$Ca_4Sb_4O_{13}F_2$	2.2	1.47

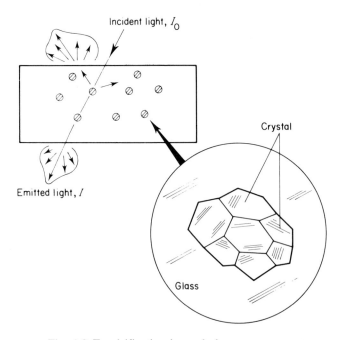

Fig. 4.3 Emulsification in opal glass

4.3 WHAT IS THE DIFFERENCE BETWEEN SILICA GLASS AND VYCOR GLASS?

Silica glass is a typical single-ingredient glass. It is composed of one type of SiO_2 and has great practical value with its low expansion rate $(4.5 \times 10^{-7} degC^{-1})$, thermal resistance $(-1,200\,°C)$, and corrosion resistance. Therefore, a high melting temperature is necessary for the manufacture of silica glass, and the crucible melting and hand blowing generally used for glass are impossible. Because of this, Vycor glass is given the same characteristics as silica glass through the addition of chemical processing after it has been melted and formed by processes like those for normal glass and shaped by glass-blowing. It contains 96–98% SiO_2, and has almost the same physical properties as silica glass (see Table 4.2). As a typical hard glass, borosilicate glass (pyrex for chemical use) is lined up too.

Table 4.2 Comparison of the properties of silica, Vycor, and pyrex glasses

	Silica	Vycor	Pyrex
Density	2.20	2.18	2.23
Knoop hardness	650	530	480
Young's modulus (kg/cm^2 × 10^5)	7.8	7.0	6.4
Poisson's ratio	0.14	0.19	0.20
Bending strength (kg/cm^2)	600	600	400–700
Linear expansion coefficient 0–300 °C	4.5×10^{-7}	8×10^{-7}	32.5×10^{-7}
Specific heat capacity (cal/g °C)25 °C	0.251	0.178	0.186
Thermal conductivity (cal/cm s °C)	3.2×10^{-3}	3.4×10^{-3}	2.6×10^{-3}
Softening point (°C)	1,650	1,500	820
Slow cooling point (°C)	1,150	910	560
Striction point (°C)	1,070	820	510

A look at the methods for making glass, beginning with melting, shows that the original raw material such as powders of SiO_2 (75%), B_3O_3 (20%), and Na_2O (5%) is mixed and then melted in a crucible at 1,500 °C. After it becomes a uniform glass, it is formed. It is possible to make hollow wares using molds in the same way as with standard glass. Fig. 4.4 shows the forming of a flask by blow molding. Next, the piece, which has been cooled, is thermally treated at 500–650 °C. A phase-separation peculiar to borosilicate glass occurs. This phase-separation is a phenomenon in which spheres about 40–250 Å across, formed from Na_2O–B_2O_3, are precipitated into the SiO_2 matrix. The Na_2O–B_2O_3 system precipitated in this way can easily be solved out by acids, and, for example, the SiO_2 frame left when it is treated with hydrochloric acid forms a sponge-like cellular glass. In this condition it is serviceable as a material in which porosity is made use of like the filters (molecular filters) made from silica glass. If this porous substance is resintered at 700–1,000 °C, a 20–40% volume shrinkage takes place and it

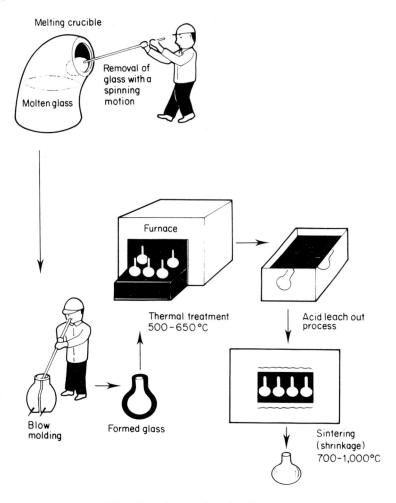

Fig. 4.4 Manufacturing method for Vycor glass

becomes a poreless, transparent glass. In this condition a 96–98% SiO_2 glass has been created, and the forming processes allow for the production of complex shapes. Before sintering, the high porosity glass shows properties similar to those of silica glass in being mechanically rigid, showing no swelling and having a large relative surface area (150–200 m^3/g), and thus it is being tested for application in biochemical and medical fields. In terms of bioceramics, a porous solid which is impregnated with glass and has a composition compatible with the living body is being investigated for applications in medical chemistry.

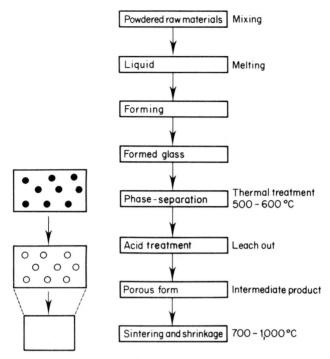

Fig. 4.5 Process and glass structure

4.4 GLASS WHICH DARKENS WHEN EXPOSED TO SUNLIGHT HAS BEEN USED IN SUNGLASSES, BUT WHAT IS ITS MAKE-UP?

Change of transmissivity (coloration, change of color, darkening) with respect to light and reversibility back to the original color or lack of color is called photochromism, and an example of its practical use is the photochromic glass used for glasses. It darkens as soon as it is exposed to sunlight, but even if it is returned to a dark room, the darkening remains, and it takes a little while for it to return to its original condition. A look at the make-up of photochromic glass can begin with the way it is made. First, the main constituent in the glass composition is boron-silicate and the light-sensitive constituents are silver halides (Ag–F, –Cl, –Br), with Cu, Cd, etc., added as sensitizers. Table 4.3 shows one example of composition. The constituents that regulate the refractive index and dispersion of the lens are PbO, BaO and ZrO_2. These constituents are mixed and then melted at 1,500 °C, and after they have formed a uniform glass, it is cooled slowly to form a glass solid. However, in this condition, the photosensitivity of the glass is too low for practical use. Fig. 4.6(a) shows the silver halides in the glass after melting.

Ions, such as Ag^+ and Cl^- exist separately within the network structure of the glass. If the glass subsequently undergoes thermal treatment in the neighborhood of 550 °C, the phase of which Ag–Cl is the main constituent separates out within the glass. Fig. 4.6(b) shows this phase-separation phenomenon. The AgCl phases produced in this way measure about 100 Å in an ideal glass, and if they are larger, there is a loss of transparency, if smaller photosensitivity is insufficient. The photosensitivity is controlled by the temperature and time schedules of the above-mentioned thermal treatment. This silver halide, which becomes the kernel of a photochemical reaction, is for the most part very sensitive to ultraviolet rays and darkens as a silver colloid.

Table 4.3 An example of the composition of photochromic glass (wt%)

SiO_2	55.9	Ag	
Al_2O_3	9.0	F	
B_2O_3	16.2	Cl	<3.0
PbO	5.1	Br	
BaO	6.7		
ZrO_2	2.3	CuO	
Na_2O	1.9	CdO	0.02
Li_2O	2.6		
Total	99.7		

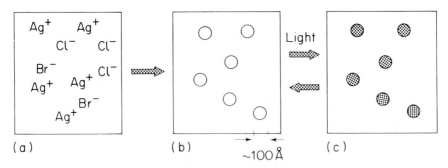

Fig. 4.6 Darkening set-up of photochromic glass

In photography, since the halogen is washed off and run out of the system after the light sensitized Ag from the exposed Ag^+X^- has been reduced to a colloid in the chemical treatment called developing, there is no recombination. In the case of photochromic glass, a spontaneous recombination takes place because of thermal oscillation when the light stimulus is removed, and the glass becomes transparent. These reactions are completely reversible, and with 100-Å-sized units in a dispersed system, a complete barrier is created by means of the stable glass matrix. Also, it remains stable over a long period, even with repeated exposure to light stimuli. Also, in terms of commercially available glass products, there is neither a detectable deterioration after coloration–decoloration tests where the reaction is repeated several tens of thousands of times, nor is there any fatigue phenomenon discernible in the glass.

In order to improve the coloring speed and photosensitivity, Cu and Cd, which function to enhance the reaction that changes the Ag ions into the Ag colloid, are added. Also, they are important constituents for controlling the size of the phase-separation mentioned above.

Fig. 4.7 shows the speed of coloration (darkening) with exposure to light and the decoloration situation (clearing) at its removal. In general, the darkening is quick, but the time needed for the glass to become clear is a problem.

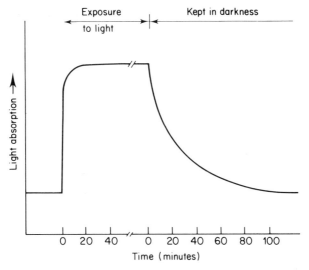

Fig. 4.7 Speed of darkening and clearing

4.5   EXPLAIN THE GLASS DOSIMETERS THAT MEASURE RADIATION

If glass is exposed to electromagnetic waves, a large or small coloration associated with a change in structure occurs. Solarization is an example of this phenomenon. In recent years, the growth of peaceful uses of X-rays and atomic power have made the management of the amount of radiation exposure more and more necessary. Along with various other dosimeters, glass dosimeters have come into use. The characteristics of glass dosimeters include the ability to precisely measure integral doses of radiation and give stable measurements during prolonged exposure. Thus, they are convenient for learning the history of exposure after the fact. When glass is exposed to radiation, color centers and radio photoluminescent centers are generated, and at the same time, radiation traces are left in the glass structure. With chemical polishing, these can be observed as etch pits. Fig. 4.8 shows the relationships between these phenomena in the glass and the measuring methods.

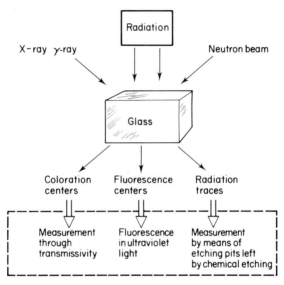

Fig. 4.8 Measurement methods and effects of radiation and glass

The color center method is used mainly for quantity measurements of X-rays and γ-rays: after the radiation exposure, the coloration within the glass is evaluated by optical absorption (that is, the opposite of transmissivity). The sensitivity and decay of color centers vary with wavelength, and as is shown in Table 4.4, most measurements are made with ultraviolet light. As for what was called decay, it is when the electron–hole pairs generated by the exposure to radiation are recombined and the color centers disappear, and it is also called fading. Through the selection of the constituents of the glass, a high linearity between the dose of radiation and the coloration may be obtained, and sensitivity may be enhanced through the selection of the wavelength used for measurement. The applicable doses and wavelengths for various types of glass are shown in Table 4.4. Cobalt glass exhibits good linearity at 400 nm over a wide range of doses, from 10^3 R to 10^7 R, and its fading is stabilized at room temperatures. Also it gives accurate data for accumulated doses over a long period of time. Repeated measurement is possible.

Table 4.4 Types and characteristics of glass dosimeters

	Dosimeter glass type	Applicable dose	Characteristics of measuring method	
Coloration center	Cobalt glass	10^3–10^7 R	400 nm	Measurement by amount of absorbed light
	Glass with manganese-iron	$\sim 3 \times 10^6$ R	540 nm	
	Magnesium-phosphate glass	10^4–10^7 R	400 nm	
	Silver activated phosphate glass	10^3–10^5 R	320 nm	
Fluorescence center	Silver activated lithium phosphate glass	~ 1 mR	Orange fluorescence stimulated by 365 nm UV light	
Others	Thermal luminescence Tb glass	0.1×10^3 R	Thermal luminescence, number of etching pits from chemical etching, not sensitive to γ-rays	
	Radiation traces phosphate glass with Th, U	neutron beams (nuclear fission of Th, U)		

The dosimeters that make use of radio photoluminescent centers have great practical value as low-level dose dosimeters, and a silver phosphate glass is used for this purpose. An irradiated piece of this glass shows an orange fluorescence with excitation by the 365 nm ultraviolet light from a mercury lamp. The construction is such that the amount of radiation is measured by the amount of this fluorescence. Contaminations, especially substances that give rise to fluorescence, must be removed from the surface of the glass. Through chopping of the ultraviolet light and other devices, it is possible to measure radiation up to 1 mR. Since the apparatus is complete, its operation is simple and repeated measurements are possible. Radio photoluminescent centers are created by silver ions. Two types of electron–hole pairs are formed in the glass by the radiation, (electron) + (Ag^+) → (Ag^0 center) and (hole) + (Ag^{2+}) → (Ag^+) center, each of which functions as a center of fluorescence. Simultaneous measurement of neutron beams is also possible.

With terbium lithium·aluminosilica glass, which makes use of thermoluminescence, it is possible to measure γ-rays of 0.1 R to 10^3 R. However, the sensitivity is inferior when compared to that of LiF.

In the glass containing ThO_2 and UO_2, the absorption of neutrons causes nuclear fission, and the traces left, etch pits, can be observed by means of chemical polishing. The number of etch pits corresponds quite well to the number of neutrons, and by using three kinds of glass with different proportions of U^{235} and ThO_2 jointly, the energy distribution of the neutrons can be known. The structural defects left in the glass, which were mentioned above, are stable, and this is a highly reliable, stable method for measurement.

4.6   WHAT KINDS OF RADIATION-PROOF GLASS ARE THERE?

Along with the many directions in which radiation is being used, such as the use of atomic energy and radioactive isotopes, radiation protection has become a big topic. There are everyday pieces of apparatus, such as color television sets, that use electron beams. Since glass is a solid as a transparent material, there is a great effort being put into devising types that protect against small amounts of radiation without loss of transparency.

The shielding capacity depends on the kind of radiation. It is determined in principle by the kinds and numbers of elements passing through the substance and by density. The shielding capacity does not depend on the bonding conditions of the glass. It is characteristic of the glass state that the capacity for self-restoration of the defects caused by irradiation is very great, and above all, since it is transparent, its greatest merit is that it can be used as an inspection window.

When an X-ray collides with a substance, a hole (positively charged hole from which the electron has been knocked out) and electron pair is created (electron–hole pair production), recoil electrons are released (photoelectric effect), and at the same time, the energy of the scattered X-ray is lost

(Compton effect). These electrons concentrate to form color centers, and as a result, browning takes place and the transparency lessens. Also after the charge has accumulated, if it is discharged at once (recombination of electrons and holes), the glass has been known to crack. Raising the electrical conductivity of the glass allows the electric charges to gradually recombine and prevents the discharge that leads to cracking. In order to decrease the color centers, it is effective to add Ce to the glass. This brings about a stable working in which Ce^{3+} ions combine with holes and Ce^{4+} ions with electrons.

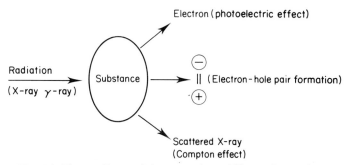

Fig. 4.9 Three effects of the operation of X-rays in a substance

In Table 4.5, there are examples of the make-up of X-ray and γ-ray shielding glasses. The general equation for transmission in any given material is

$$I = I_0 e^{-\mu t} \tag{1}$$

Here, I_0 is the incident radiation, I the transmitted radiation; μ is called the absorption coefficient, $\mu = \omega\rho$. ω is the mass absorption coefficient, ρ the density. On the other hand, if the dose of the secondary radiation is considered, a build-up coefficient is added,

$$I = B I_0 e^{-\mu t} \tag{2}$$

and more general changes can be understood. Here, the build-up coefficient B is defined as $B =$ (the total radiation reaching the detector)/(the primary radiation reaching the detector).

Absorption of neutron beams shows high selectivity of elements, and water, Gd, Eu, Cd, B^{10}, Dy, etc., have strong absorption effects. By making glass that contains large amounts of these elements, the shielding effect can be enhanced. Glass composed of CdO 63.8%, B_2O_3 31.1%, Ti_2O 2.0% and ZrO_2 3.1% (by weight) is used for neutron shielding.

A block with a total thickness of 1 meter made up of a combination of these glasses is incorporated into the inspection windows of the hot cells (hot cave) for radiation work. A comparatively thick γ-ray-proof glass which is resistant to coloration is used on the side near the source of the radiation, and a high-lead-content glass with a large shielding effect is used on the side near the observer. Since many of these protective glasses have high refractive indices, the field of vision from the window is wide.

Table 4.5 Types of X-ray- and γ-ray-proof glass

Density	Composition (wt%)				
3.2–3.6 Lead glass	PbO 42.5	SiO$_2$ 42.5	K$_2$O 5.5	Na$_2$O 8.0	Ce 0.9–2.5
4.2–5.2 High lead glass (with Ce)	SiO$_2$ 29	PbO 62	BaO 8.1	CeO$_2$ 0.9	
2.5 Soda-lime glass (with Ce)	SiO$_2$ 72	Na$_2$O 15	Al$_2$O$_3$ 1.0	CaO 11.6	CeO$_2$ 1.4
6.2 High lead glass	SiO$_2$ 27	PbO 71	K$_2$O 2.0		

Lead equivalent: Expressed by the thickness of lead with the same absorption coefficient as the glass. For example, glass with a lead equivalent of 0.274 and a thickness of 1 cm has the absorption of 0.274 cm of lead.

4.7 WHAT IS THE GLASS THAT IS USED IN THE TREATMENT OF RADIOACTIVE WASTE?

After the enormous amount of heat generated during the fission of uranium has been used for the generation of electricity, most of the new elements produced when the uranium is split in two are radioactive. These are called fission products, and their amount increased with the amount of power generated. The spent nuclear fuel is kept sealed in its strong zircaloy or stainless steel sheath and stored in a pool until the unused uranium and newly generated plutonium are separated out during a reprocessing cycle as either solids or liquids. During the reprocessing cycle, all the elements are separated out according to their chemical properties by a solution extraction process as nitrate solutions. At this stage, almost all of the fission products are extracted as water-based nitrate solutions. This can be called highly radioactive liquid waste, and it contains many elements (almost all of those in the periodic table) such as Cs^{137}, Sr^{90}, etc. Also, since these elements actively undergo repeated radioactive decay, the radiation and heat from this decay must be dealt with. Storage in the form of a solution seems somewhat inconvenient, and should the tanks be holed, there is the danger of leakage. Therefore, because of the need to convert the waste to a solid form for storage, making it into glass has been considered.

The glass state is different from the crystal state in that all of the components lack crystal affinity and are melted together to form a glass, making for high mixability. Nevertheless, the formation of a homogeneous glass is limited by the various elements included. In particular, sodium and molybdenum coexist among the elements that are part of the radioactive liquid waste, and this makes conversion to glass difficult. In order to make the glass conversion easy, additives are being devised, and in terms of the glass, studies along the line of improving the multiconstituent nature are being carried out. As a result, borate glass has been selected in consideration of stability, water resistance, thermal conductivity and ease of manufacture. In Japan, these glasses are being studied by the Power Reactor–Nuclear Fuel Development Corp. An example is shown in Table 4.6.

Table 4.6 Example of highly radioactive solid glass waste (Japan)

Role	Constituent	(wt%)
GA	SiO_2	43.0
	B_2O_3	14.0
	Al_2O_3	4.0
	Li_2O	3.0
	Na_2O	10.0
	K_2O	2.0
	CaO	2.0
	ZnO	2.0
PI	CoO	0.16
	Cr_2O_3	0.18
	Fe_2O_3	2.12
	NiO	0.60
FP	Rb_2O	0.22
	SrO	0.64
	Y_2O_3	0.42
	ZrO_2	2.87
	MoO_3	3.16
	Cs_2O	1.60
	BaO	1.04
	La_2O_3	0.94
	Ce_2O_3	1.80
	Nd_2O_3	4.44
	U, Pu etc.	small amounts

GA: Glass additives.
PI: Inert elements in the process.
FP: Fission products.

Fig. 4.10 shows the process of conversion of the highly radioactive liquid into a solid glass radioactive waste. In order to improve the solubility of the water solution, the waste is progressively treated since the materials are all nitrates. The denitration–concentration process is the stage in which the nitrates are decomposed and the water is evaporated to make the solution into a slurry. Next, the additives for the conversion to glass are added, and a uniform glass is produced at 1,200 °C in a glass-melting furnace. When it is sufficiently melted, it is poured into a metal canister. Once the column-like waste solids are formed, they are stored carefully in forced-air- or water-cooled facilities. Radioactive decay continues in these glassified nuclear fission products, and it is necessary to maintain sufficient surveillance over the heat and radiation produced. All of these processes must be manipulated through remote control in the concrete cells that protect against radiation. Based on this technology, completely unattended glass-melting operations are being established, and it can be assumed that these will give a stimulus to the glass industry as a whole.

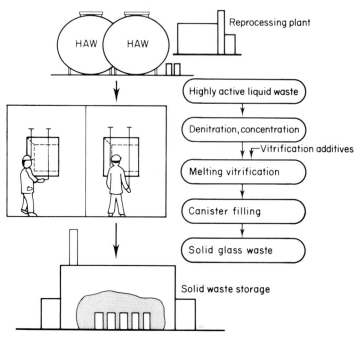

Fig. 4.10 Process for turning highly active liquid waste into solid glass waste

4.8   WHAT KIND OF GLASS IS LASER GLASS?

In order to produce a laser action in a solid, it is necessary to select a stable solid which includes elements such as Nd^{3+}, Cr^{3+}, Mn^{3+}, Yb^{3+}, Dy^{3+}, etc., which are highly efficient for inducing an emission when stimulated by external light. If one considers glass as a solid solvent, it is ideal as a matrix in which these elements can be distributed by melting. The uses of solid lasers are for high-speed repetition and high output, and there are especially high expectations for the realization of nuclear fusion through inertial confinement with glass lasers. Among them, the glass laser containing Nd^{3+} is the closest to becoming practicable.

If a glass rod containing 3–8 wt% Nd^{3+} is exposed to a Xe lamp, the energy absorbed by the Nd^{3+} ions is stored. This is called optical pumping; energy is supplied and the excited electrons wait for a chance to fall back to the W_1 state from the W_2 state to which they have been raised, and on returning

Table 4.7 Characteristics of laser glass (from HOYA glass data)

Laser characteristic	LSG-91H	LHG-5	LHG-7	LHG-8	LHG-10
Nd_2O_3 (wt%)	3.1	3.3	3.4	3.0	2.4
Nd^{3+} ion concentration (10^{20} ions/cm³)	3.0	3.2	3.1	3.1	3.1
Area of emission cross-section σ_p (10^{-20} cm²)	2.7	4.1	3.8	4.2	2.7
Duration of fluorescence (μs)	300	290	305	315	384
Gain coefficient (cm⁻¹/J/cm³)	0.144	0.217	0.202	0.223	0.143
Half value width (290 K) $\Delta\lambda$ (Å)	274	220	222	218	265
Median wavelength λ_p (nm)	1,062	1,054	1,054	1,054	1,051
Damping coefficient (1,054 nm)(m⁻¹)	0.1 (1,062 nm)	0.123	0.131	0.1	0.15
Probe efficiency (%) (10 mm dia. × 160 mm, R = 80%)	1.15	1.83	1.82	1.83	
Oscillation threshold value (J) (10 mm dia. × 160 mm, R = 80%)	52	40	35	32	
Relative capacity index σ/n_2	1.00	1.87	2.12	2.17	

generate a fluorescence. This light reflects back and forth along the length of
the rod from mirrors affixed to each end and resonates with the light emitted
from each Nd^{3+} ion to become a single wave and pass through one of the
mirrors. The wavelength of this Nd^{3+} glass laser is 1.06 μm.

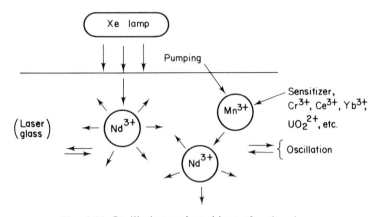

Fig. 4.11 Oscillation and workings of a glass laser

In order to enhance the efficiency of the laser, the following requirements are made of the matrix. The gain is expressed as the product of
capacity for emitting fluorescence (area of the induced emission cross-
section σ) and the amount of energy stored (inversion density ΔN), for
each of which it is desirable to have a large value. When ultra-high power
(300 TW (terawatt = 10^{12} watt) in the case of nuclear fusion) is required,
the method is to line up glass lasers lengthwise for successive amplification
of the output. For an oscillator, it is necessary that σ be large, and for an
amplifier, that ΔN be large. In order for the laser action to be stable, it is
necessary that the refractive index varies little with the electric field
(non-linear constant n_2, $n = n_0 + n_2|E|^2$, E = electric field) and that there
is little change of temperature along the optical path. The optical path is
expressed as the sum of the refractive index and linear expansion, and in a
glass it is regulated by the addition and subtraction of minute amounts of
additives ($dS/dT = dn/dT + (n - 1)\,\alpha$, n = refractive index, α = linear expansion coefficient). Also, if local variations in temperature give rise to
changes in the refractive index, striction is generated, the oscillation is
disrupted, and there is a factor that disturbs the phases. Also, if the glass
matrix contains foreign elements such as platinum, heat is generated in
that area and is a factor that leads to breakage. Fe, Cu, Ni, Co, V, Pr,
Dy, Sm are impurities that interfere by absorbing light and should be
excluded. Platinum crucibles are used for melting high purity glass, but a
process to keep small platinum particles from mixing in alleviates any
worry. Based on the above requirements, sodium carbide, borate, alkali-
barium, and barium crown glass systems have been developed, and

furthermore, for high-power Nd^{3+} glass lasers, phosphate and phosphate-fluoride glasses seem promising. In these glasses, a composition in which the gain coefficients are high, the non-linear refractive indices are low compared with silicates and borates, and the coefficients of thermal change along the optical path (dS/dT) are almost zero is obtained. Changes in constituents according to the oscillator or amplifier and shapes ranging from rods to wafers are conceivable in terms of usage and design.

4.9 HOW IS THE COLORED GLASS THAT IS USED IN PHOTOGRAPHIC FILTERS, ETC., MADE?

The filters used for photography are either glass in which the glass itself is colored or vaporization films applied to transparent glass. It has long been known that glasses containing ions of transition metals take on colors. The cobalt blue used in the well-known stained glass called Chartres blue, has been used for glass craftwork for a long time. In this way, glass is colored by additives, the two large divisions of which are:

colloidal coloring: gold (red), copper (red), silver (yellow-brown), selenium (red), etc.
ionic coloring: copper (blue-green), iron (dark green–yellow-green), manganese (purplish red), nickel (purple), cobalt (blue), chromium (green), uranium (green), carbon (brown),

and the tones of these colors are changed by the components of the glass, especially alkalis. Most glasses with colloidal coloration are colorless or slightly colored at the time of melting, but brilliant coloration may be obtained through a thermal treatment coordinated with colloidal growth. This operation is called striking. When additives that show remarkably little absorption in the visible spectrum but absorb ultraviolet rays are included, an UV cut filter is produced. These sharp cut filters extend to the visible spectrum, and some typical examples of filter usages are shown in Table 4.8.

One of the methods for expressing color tones exactly is to show a plot of spectral transmissivity vs. visibility on x–y coordinates. The method known as a CIE chromaticity diagram is shown in Fig. 4.12 (CIE = Commission International de l'Eclairage 1931, JIS Z8701; 1952).

Table 4.8 Uses of sharp cut filters

L39	Colorless	Cuts out ultraviolet, from 390 nm
Y44	Light yellow	Cuts out ultraviolet, from 440 nm
Y48	Yellow	Cuts out ultraviolet, from 480 nm
Y52	Dark yellow	Cuts out from blue (520 nm) to ultraviolet
O56	Yellowish red	Cuts out short wavelength side from 560 nm, for microphotographic use
R60	Red	For primary color analysis, red; infrared photography

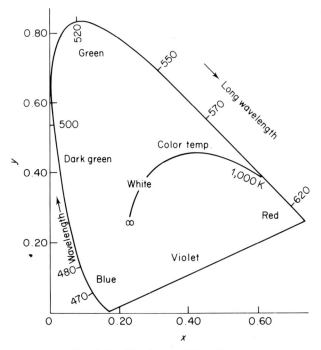

Fig. 4.12 CIE chromaticity diagram

Color tones are primarily expressed in color temperatures which make use of the convenience of a choice of light sources and indicate the light which corresponds to the spectral distribution of the radiated light from a black body at that temperature. Three types of light sources that are commonly used for photography are:

light source A: a gas-filled tungsten bulb with a color temperature of 2,854 K.

light source B: daylight at noon with a color temperature of 4,870 K. When light source A is filtered it equals light source B.

light source C: corresponds to the light in daytime under blue sky, about 6,500 K.

Also there is a light source E in which the energy of the light is almost the same for every wavelength in the visible spectrum. It has a color temperature of 5,270 K which is close to black body radiation. Light from source A can be converted into any of the others, B, C, E, by filters.

Next, there are interference filters which transmit light selectively in amounts equal to the thickness of an interference film applied by vaporization, etc., to the surface. If these filters are combined with sharp cut filters, they can be used to selectively pick out narrow wavelengths (from a spectrum with many wavelengths), and transmit one.

4.10 WHAT KIND OF GLASS ARE THE FIBERS THAT ARE USED IN COMPOSITE MATERIALS LIKE GLASS-FIBER-REINFORCED PLASTICS?

Among the glass-fiber-reinforced plastics, there are thermosetting resins or FRP (fiber-reinforced plastics) and thermoplastic resins or FRTP (fiber-reinforced thermoplastics), and these are typical composite materials of increasing importance. These glass fibers are divided into long fibers and short fibers, and typical compositions for each of them are shown in Table 4.9. E glass was developed for electrical applications and C glass for chemical ones. A glass has the same composition as commercially available window glass. The composition of short fibers changes according to the way they are made. A cotton-like form is used for insulation. Silica fiber is suited for industrial type high temperature insulation and is also used for the heat-resistant tiles on the outside of the space shuttles. With the exception of silica wool, all are produced by being melted to a uniform liquid in a melting furnace, then formed into 15–30 mm diameter marbles, remelted over bushings with small holes in the bottoms and removed downwards as fibers. Long fibers can be rolled up at this point, and short fibers can be steam blown and collected in a mat. There is also a direct fiber-making process that does not pass through the marble stage.

The tensile strength of a fiber depends on the diameter, and the smaller the diameter is, the larger the value becomes, with saturation at several micrometers and 350–400 kg/mm^2. The tensile strength of the glass is remarkably affected by surface scratches of several micrometers or so. These scratches are produced just by exposure to air, and if they are handled, mechanical strength also decreases. A tensile strength of 400 kg/mm^2 should give 3g at 3μm, 30g at 10 μm, but observations show 11–25g at 10μm (140 to 320 kg/mm^2). Since a surface treatment causes the binding strength between the surface protection and the resin to increase, it has been confirmed that coating with complex chromium compounds or *silane* compounds (chloro-*silane* systems, alcho-oxy-*silane* systems, amino-*silane* systems, epoxy *silane* system resins, etc.) improves the strength 1.4 to 2.0 times when compared with untreated FRP.

As an example, with a glass fiber buried in epoxy resin, the strength of the FRP becomes greater as the amount of glass fiber is increased. The orientation of the fibers plays a role, and with a 50 vol% content arranged along the tensile direction, the strength is about 130 kg/mm^2, and for a random arrangement 70 kg/mm^2.

There are great expectations in the near future for FRTP, which can be injection molded, as an engineering plastic capable of precision forming. Nylon, polyethylene, polypropylene, PVC, polycarbonate and AS and ABS resins, etc., are mixed with pre-cut short glass fibers and supplied in a pellet form which is used as a direct charge for a molding machine and allows for injection molding.

Table 4.9 Composition of fibers for reinforced plastics

	SiO_2	Al_2O_3	Fe_2O_3	CaO	MgO	R_2O	B_2O_3	Li_2O	BaO	ZnO	
E glass	53.5	15.0	0.2	17.5	4.5	0.4	8.5	—	—	—	long
E glass	53.2	14.6	0.2	21.1	0.5	0.8	9.0	—	0.3	—	fibers
C glass	66.9	4.1	0.2	6.5	2.7	11.4	4.0	0.6	—	3.5	
A glass	72.4	1.6	0.1	7.5	4.5	13.2	—	—	—	—	
Flame method	68.6	2.2	0.1	6.9	2.4	17.7	3.7	—	—	—	short
Rotary method	60.8	3.5	0.1	8.0	3.0	15.6	6.0	—	2.5	—	fibers
Silica wool	99.1	—	—	—	—	0.7	—	—	—	—	

Table 4.10 High-elasticity glass fibers

	SiO_2	Al_2O_3	Fe_2O_3	CaO	MgO	Li_2O	BeO	TiO_2	ZrO_2	Young's modulus
Houze Hi-Mod	53.8	38.9	0.7	—	—	—	—	—	—	11,690
Imperial N-672	53.0	—	—	16.0	11.0	3.0	8.0	4.0	5.0	12,110

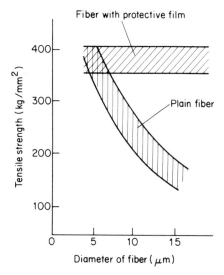

Fig. 4.13 Tensile strength vs. diameter for glass fibers. The effects of a protective film prevent a loss of strength

In a way similar to that of FRP, there are examples of portland cement and plaster reinforced with glass fiber.

For this purpose a glass containing zirconia has been developed in order to combat the alkali of the cement. The glass used to reinforce cement has an SiO_2 65 to 75, ZrO_2 7 to 11, and R_2O 13 to 23 wt% composition. There have also been tests of glass-fiber reinforcements for metal and rubber, but none have come into practical use.

The elastic constants of E glass and C glass are 7,500 and 7,000 respectively, but the high-elasticity glasses in Table 4.10, Houze 11,690 kg/mm² and Imperial 12,110 kg/mm², are being investigated to determine if they are useful for high-quality FRP devices for use in outer space.

4.11 EXPLAIN THE GLASS FIBERS USED IN OPTICAL TELECOMMUNICATIONS

Since glass is transparent and can be stretched into long fibers it is quite convenient for use as an optical telecommunications medium. In general, the essentials are the construction of a transparent medium. a part of which has a high refractive index. To make this easy to understand, here is an explanation of the concept of optical confinement through the condition of total reflection. Fig. 4.14(a) shows the situation when light enters toward the interface of two transparent materials with different refractive indices. When a ray of light moves from a medium with a higher refractive index (n_1) to a medium with a lower one (n_2), the angle of emergence r is larger than that of incidence. This relationship is expressed by $n_1\sin i = n_2\sin r$. Also ray (1) is divided into the reflected ray (1)′ and the refracted ray (1)″ at the interface. If the angle of incidence is increased from (1), at (2) the angle of emergence r_2 becomes 90°. The angle of incidence at this point, i_2, is called the critical angle and is given by $i_c = i_2 = \sin^{-1}(n_1/n_2)$. With angles of incidence greater than i_c, almost all of the light is reflected (total reflection). In this way, the light has the property of remaining in the medium with the higher refractive index. As shown in Fig. 4.14(b), the optical telecommunication fiber is a long fiber with a high refractive index that forms the inner of two coaxial layers. Light with an angle of the width of Θ along the central axis of the fiber which is introduced into the fiber shows total reflection and is transmitted. The sine of the largest value for this angle, Θ_c, is called the numerical aperture (NA). Assuming that n_1 is the refractive index of the core the light is transmitted through and n_2 is the refractive index of the outer enveloping glass (cladding), NA is shown by

$$NA = \sin \Theta_c = \sqrt{n_1^2 - n_2^2}$$

When the condition $\sin \Theta \leqslant NA$ is satisfied, the light is transmitted along the fiber. This set-up is called a step index (SI) type and is so named because the refractive indices are distributed in a step-like fashion.

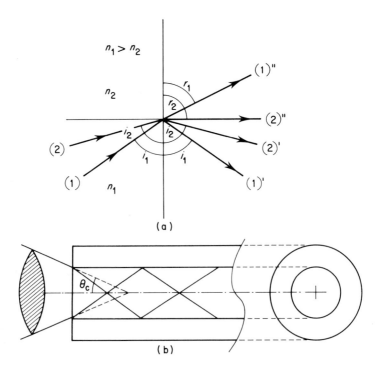

(a)

(b)

Fig. 4.14 Set-up for optical fiber transmission. (a) Refractive index and reflection. (b) Fiber cross-section

Another important type of optical fiber is the graded index (GI) type, which has the glass with the highest refractive index along the central axis with the refractive index decreasing parabolically toward the outer perimeter. In this fiber, light travels faster in the areas of lower refractive index, and since it converges on the central axis as it travels, there are almost no variations in the path of light transmitted along the length of the fiber. In the SI type, variations in the path equal to the geometric distance only are caused by the light travelling parallel to the axis and the light forming an angle with the axis and progressing by repeated reflections. If a pulse-type optical signal

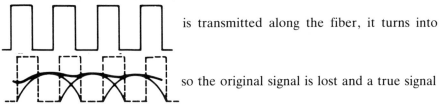 is transmitted along the fiber, it turns into

so the original signal is lost and a true signal cannot be transmitted. The distance at which the shape of the wave can be discerned is the limit of the telecommunication. And with the GI type, research is being done to find a glass in which the proportional distribution of refractive indices (a nearly parabolic distribution) is in the glass itself. The limits of the transmission signal are called the band of the fiber, and this is expressed as the product of the switching frequency of the light and the length of the fiber. For the SI type, it is generally 10–40 MHz km, and for the GI type, 200–1,000 MHz km. In Fig. 4.15, an example of an optical cable is shown. The structure is designed to protect the glass.

4.12 WHY DO OPTICAL TELECOMMUNICATIONS FIBERS, WHICH ARE TRANSPARENT, USE INFRARED LIGHT?

The basis of optical telecommunications is transmitting a signal by the propagation of light. By confining the optical signal that is being transmitted within a glass fiber, the light, which originally has the property of travelling in a straight line, is bent freely and can be transmitted over long distances, the main point of the optical signal. All substances give rise to absorption and scattering of light. For the efficiency of propagation, it can be said that a vacuum is best. If one thinks about propagation in the air, the question of why the sky is blue must be answered. The fluctuation of the components of air cause light-scattering. This phenomenon was discovered by Lord Rayleigh and is called Rayleigh scattering, and is inversely proportional to the fourth power of the wavelength of the light. Therefore, the shorter the wavelength of the light, the greater the scattering is, and longer wavelengths are transmitted easily. When light is viewed from a perpendicular to the path of the light, the amount of short wavelengths increases; therefore, more blue is scattered, and the sky looks blue. Also, the fact that the sunset is red is evidence that longer wavelengths are easily transmitted.

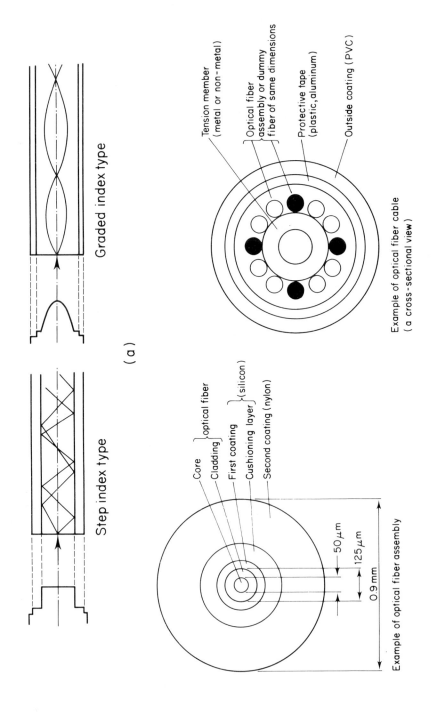

Step index type

Graded index type

(a)

Core ⎱ optical fiber
Cladding ⎰
First coating ⎱ (silicon)
Cushioning layer ⎰
Second coating (nylon)

50 μm
125 μm
0.9 mm

Example of optical fiber assembly

Tension member (metal or non-metal)
Optical fiber ⎱ assembly or dummy
fiber of same dimensions
Protective tape (plastic, aluminum)
Outside coating (PVC)

Example of optical fiber cable
(a cross-sectional view)

Fig. 4.15 The types of optical fibers and an example of an optical cable

The most important factor in preventing the propagation of light in the longer wavelengths is the absorption of light due to water impurity. The resonant frequency of the expansion and contraction of H–O–H is at 2.8 μm, and the overtones also give rise to absorption. Correspondingly, transmission in air is not a very efficient transmission. Of course, the physical obstacle of non-transparent substances must be considered. Besides, it is necessary to have a light source with remarkably high parallelism. Based on the concept of cavity resonators, optical cables have been tested, but an appropriate material has not been found. At present optical communication fibers composed mainly of silica glass are used for typical optical telecommunications media. With the progress in research on optical fibers, the transparency of silica glass has almost reached the theoretical limit of 0.2 dB/km for a wavelength of 1.6 μm, and this is as of 1980. The profile of transmission loss shows that the baseline of resonant absorption extends from the long wavelengths, and is coupled to a valley in the line drawn from the short wavelength side. Furthermore materials that have valleys in the long wavelengths are being studied, and glasses with fluoride content have a low loss area from 3 to 4 μm. A substance which transmits long wavelengths and has a loss limit of 10^{-4} dB/km has been found.

Impurities that bring about transmission losses are the transition elements like Fe, Ni, Cr, Co, Mn, etc., that show up in coloring. These must be removed from the glass as absolutely as possible. Highly pure silica is produced as the material for silica glass by repeated distillation–purification of $SiCl_4$, followed by the creation of fine particles of SiO_2 by gas phase decomposition and oxidation, and vaporization.

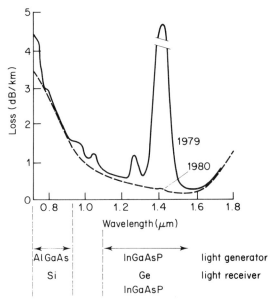

Fig. 4.16 Transmission loss in optical telecommunications fibers (dashed line is the theoretical limit)

Ge, P, etc., are used as dopants to increase the refractive index, and for decreasing the refractive index of the cladding there is B doping. All of these are converted to chlorides and undergo gas phase reactions like $SiCl_4$, decomposition and oxidation, and the pre-fiber form of silica is made.

An example of the semiconductor laser used for the light source is shown in Fig. 4.17. On a GaAs substrate a p–n pair is formed by varying the ratio Al/Ga, and between the pair, an optically active layer is formed. The light excited comes out in the direction of the arrow and is connected to the optical fiber. In order to convert the wavelength of the generated light into a longer wavelength, an In-Ga-As-P system is formed in the active layer, and a 1.1–1.6 μm laser diode is made.

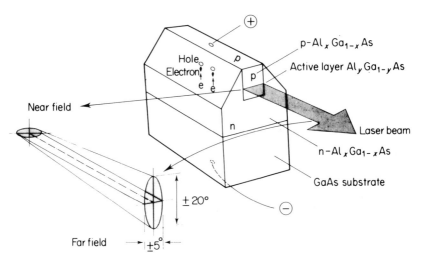

Fig. 4.17 GaAlAs semiconductor laser

Usually, the magnitude of the distribution of emitted light varies in length and width. In the connection with the optical fiber, the distribution characteristics of the diode should be taken into account.

4.13 WHAT ARE THE CHARACTERISTICS OF CRYSTALLIZED GLASS?

Properly speaking, glass is not crystalline. In the strict definition, glass equals non-crystalline. Even so, what is called crystallized or devitrified glass is made by precipitation of crystals through thermal treatment after glass has been made in the classic way into, for example, containers. In the same way concrete is made from cement and gravel, crystallized glass is made up of the glass matrix and crystals (Fig. 4.18).

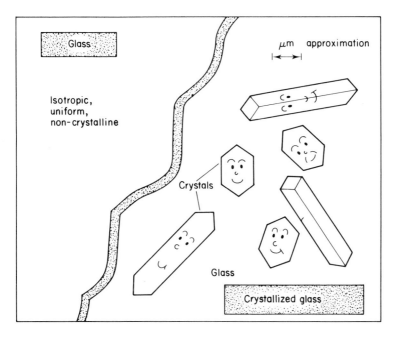

Fig. 4.18 The structure of glass and crystallized glass. With a polarization microscope the difference is clear

(1) Crystallized glass has high mechanical strength. If a crack appears in a piece of glass, the stress is easily concentrated at the ends of the crack, and breakage proceeds until the stress is released. If crystallization has taken place, the stress from the crack will be released where there is a crystallite, and the progress will be checked. The crystallized glass that is in practical use has two or more times the strength of glass in the normal glass state because of the thermal crystallization treatment. Furthermore, the 100 μm thick compressed striction layer that is set up in accordance with ion exchange method attains a value of 4,000 kg/cm^2.

(2) Thermal resistance is improved. The characteristics of crystallized glass depend on the material precipitated out as crystals, but if β-silica is crystallized out, it has an axis with a negative thermal expansion coefficient and the thermal expansion coefficient of crystal glass with this composition can be taken as 0. Like silica glass (thermal expansion coefficient 4.5×10^{-7}degC^{-1}) this crystal glass can withstand rapid heating and cooling. Since the crystalline parts remain solid at higher temperatures than the unmodified glass (have a higher melting point), crystallized glass remains unsoftened up to higher temperatures. In essence this complex-phase glass has greater high temperature viscosity.

(3) It transmits light; that is, there is also transparent crystal glass. The formation of crystals from glass was called devitrification and was considered

undesirable in glass production, but in crystallized glass, if the size of the crystallites is uniformly less than the wavelengths of light (of the order of 0.1 μm), or if the crystals are isotropic and their refractive index is equal to that of the glass matrix, transparent crystallized glass may be obtained. It is used for the combustion tubes of heating stoves.

(4) There are crystallized glasses with unique electrical characteristics. Examples of crystallized glass in which electrical conductivity, dielectric constants, magnetic characteristics, and electro-optical characteristics are especially excellent, such as $BaTiO_3$, cordierite, ferrite, $NaNbO_3$, $(Ba\cdot Sr\cdot Pb)\cdot Nb_2O_6$, etc., are being investigated. In Table 4.11, $NaNbO_3$ crystallized glass is given as an example. In the case of a crystallized glass that has a 70% by volume precipitation of $NaNbO_3$ crystals that are several hundred angstroms in diameter in a solid solution of Cd, optical anisotropy is generated in proportion to the size of the electric field and retardation is observed. Possible practical applications include devices for electrical control of light, optical switches and devices for display of color etc. Electro-optical devices made from glass have the characteristics of uniformity and ease of mass production.

Table 4.11 $NaNbO_3$ crystal glass

Composition	Nb_2O_5 66, Na_2O 15, SiO_2 14, CdO 3, TiO_2 2
Precipitated crystal	$NaNbO_3$: Cd
Crystal diameter	Several 100 Å
Amount of crystal precipitation	70 vol%

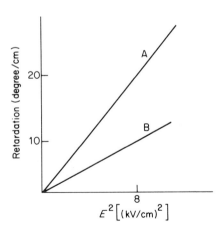

Fig. 4.19 Unusual nature of $NaNbO_3$ crystallized glass in an electric field. Dielectric constant A: 540, B: 356

5 New Technology of Ceramics: Questions and Answers

Chapter 5 is concerned with the new technology of ceramics. First, new process technologies such as ultra-fine particle powders, ceramic laminations and multilayer technology are taken up. Next, materials which are common topics, like ceramic surgical replacements, superconductors, etc., are taken up. Also, there are amorphous ceramics, which are interesting as ceramics of the future. Since there are ceramics with unique properties such as high strength, high toughness, high thermal conductivity, etc., these will be touched upon. Furthermore, the new technology of bioglass will be dealt with.

5.1 WHAT KINDS OF CERAMICS ARE USED IN THE LIVING BODY?

When the body's bone, muscle, and other tissues, or heart, kidneys, and other organs fail to function normally because of obstacles like disease, injury, deformity, and aging, repair or replacement with artificial materials has been attempted for some time. Putting the soft structures and organs aside, it is natural to think of ceramics for hard materials such as bones, but until now only metals and plastics have been used; ceramics have not been used for anything but teeth. This was because ceramics had some bad points. But these ceramics have undergone recent improvements and they are coming into use for various tissues and organs, especially, hard tissues like bone, etc.

The following can be thought of as uses for ceramic artificial bones: (i) long bones, (ii) joints, (iii) partial jawbone replacements, (iv) replacements for the roots of teeth, (v) replacements for bone deficiencies, (vi) plates and screws for securing broken bones, (vii) tendons and ligaments, and (viii) heart valve replacements.

In order to use ceramics in the body, the following conditions must be satisfied:

(a) no effects that are harmful to the body, such as toxicity, irritation of tissues, carcinogenesis, or thrombosis;
(b) compatibility with living tissue;
(c) strong adherence to the surrounding bones and tissues, if possible, chemical bonding ideal;
(d) tensile, bending, compression, shearing, etc., strengths greater than natural bone and little fatigue (phenomenon of growing weak over time) while in body fluids;
(e) wear resistance;
(f) hardness and elastic coefficient roughly equal to those of natural bone;
(g) easy forming and processing.

It can be said that conditions (a) through (c) are biological, (d) through (f) are mechanical, and (g) makes handling easy.

In Table 5.1, the main ceramics that have been investigated for biological uses are shown. In this table, the materials are divided into the two categories of bio-inert and bio-active. Bio-inert materials produce almost no changes in the living body, and they include stable compounds like C, Al_2O_3, Si_3N_4, TiO_2, as well as CaO–Al_2O_3 and Na_2O–Al_2O_3–SiO_2 systems. Bio-active materials cause decomposition, absorption, reactions, and precipitation, and they include both CaO and P_2O_5.

Since ceramics have excellent compatibility with the living body, they have been used for the repair and replacement of bones, tooth roots, heart valves, etc. Examples of these are shown in Fig. 5.1. With the development of this field, one can expect that excellent materials that reduce the patient's pain will become practicable.

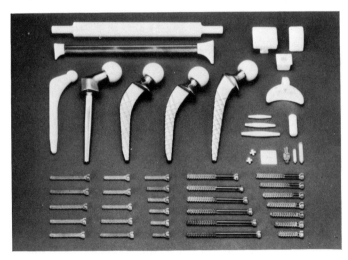

Fig. 5.1 Examples of repair and replacement materials (courtesy of Kyocera)

Table 5.1 Main ceramics for the living body

Bio-inert materials	Carbon (C): thermally decomposed carbon, glassy carbon, carbon fiber Alumina (Al_2O_3): single crystals, polycrystalline products, high-porosity products Others: polycrystalline Si_3N_4, TiO_2, $CaO–Al_2O_3$, high-porosity $Na_2O–Al_2O_3–SiO_2$ systems
Bio-active materials	Glass containing CaO and P_2O_5: $Na_2O–CaO–SiO_2–P_2O_5$ systems Crystal glass containing CaO and P_2O_5: $Na_2O–K_2O–MgO–CaO–SiO_2–P_2O_5$ systems, $MgO–CaO–SiO_2–P_2O_5$ systems Hydroxyl apatite ($Ca_{10}(PO_4)_6(OH)_2$): polycrystalline products, granules High-porosity $3CaO·P_2O_5$ Others: $CaO–Al_2O_3–P_2O_5$ systems, $K_2O–CaO–Al_2O_3–SiO_2–P_2O_5$ system glass fibers

5.2 WHAT APPLICATIONS ARE THERE FOR HIGH-STRENGTH, HIGH-TOUGHNESS CERAMICS?

Ceramics are brittle materials, and since a slight impact can fracture them, this is pointed out as a drawback when they are used as structural materials. Various studies aimed at improving this type of property in ceramics are continuing. In this case it is necessary to distinguish between 'strength,' which involves static stress, and 'toughness,' which involves dynamic stress.

Attempts to improve the resistance to shock (increase toughness) in most common materials which have been available for a long time have been carried out by means of increasing the strength. These materials are alumina, silicon nitride, and silicon carbide. For example, silicon nitride with a bending strength of 100 to 130 kg/mm^2 has recently been made, but the expected toughness results were not obtained. While these studies were going on, several interesting reports on the utilization of secondary phases emerged. Among these, the mechanical properties of partially stabilized zirconia were investigated, and ever since the tetrahedral/monoclinic martensite transformation was used for the absorption of destructive energy, the development of high-toughness ceramics that make use of what is called the stress-induced phase-transformation (SIPT) or PZT has been worthy of notice. (For the details on zirconia, see Section 2.6.)

The numerical value that expresses the toughness is the fracture dynamics parameter K_1. This indicates the strain and stress conditions at the point of the crack, and K_{1c}, especially, indicates the critical value at which sudden growth of the crack will take place. The toughness of various ceramics reported recently are shown in Table 5.2. It is worth noting that the values of $K_{1c} \doteqdot 6$ to 10 in two-phase ceramics such as PSZ–CaO, PSZ–Y$_2$O$_3$, and ZrO$_2$–Al$_2$O$_3$ are far superior to those of Si$_3$N$_4$ and SiC which have been developed as structural ceramics up to now.

Table 5.2 Fracture toughness of ceramics

Materials	K_{1c} (MN m$^{-2/3}$)
ZrO$_2$–Y$_2$O$_3$	6–9
ZrO$_2$–CaO	9.6
ZrO$_2$–MgO	5.7
ZrO$_2$	1.1
ZrO$_2$–Al$_2$O$_3$	9.8
Si$_3$N$_4$	4.8–5.8
SiC	3.4
B$_4$C	6.0
Al$_2$O$_3$	4.5
Single-crystal spinel	1.3

The high-toughness mechanism is the previously mentioned tetrahedral transformation. When a tetrahedral crystal is transformed into a monoclinic crystal at the point of cracking, the direction of growth is redirected in various different directions and so is the energy of the crack. As a result, since most of the cracks in ceramics grow in a zigzag or branching fashion, the energy necessary for fracture is greatly increased and breakage becomes difficult.

Since zirconia shows a great improvement in the weakest point of ceramics, 'brittleness,' there are great expectations for the use of this high-strength, high-toughness ceramics. The currently known uses are (i) edged tools and other tools (e.g. cutting tools, cutters, knives, etc.), which employ its toughness, strength, and hardness; (ii) ball mills, which make use of its wear resistance, toughness, and relative density; and (iii) dies, nozzles, and sliding parts, which employ its hardness, toughness, and low friction coefficient. Some of these products are shown in Fig. 5.2.

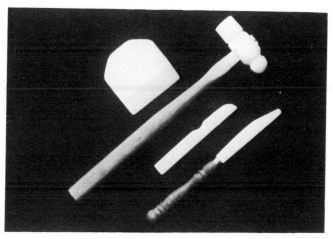

Fig. 5.2 High toughness zirconia products (courtesy of Toshiba Ceramics, Inc.)

5.3 WHAT ARE AMORPHOUS CERAMICS?

Nowadays the word 'amorphous' is often heard in terms like amorphous metals, amorphous semicoductors, and amorphous magnetic materials. From the point of view of the study of ceramics, it is natural that there should be amorphous ceramics, but the term is not so common at present. However, with a little thought, one realizes that 'amorphous' is a word that means 'glass'; therefore, amorphous ceramics is the original member of the whole family of amorphous materials.

Amorphous materials are solids that are not crystalline; that is, it can be said that molecules that form structures other than crystals are in an irregular non-crystalline configuration. Plate glass and bottle glass are amorphous and so are the non-crystalline materials obtained by sputtering Si and Ge. In the amorphous state, where atoms are not restricted to lattice points as in crystals, there is freedom in the structure, and this creates the basis for the characteristics of functional amorphous materials.

Amorphous ceramics are manufactured by sputtering, CVD, and other methods, but there is also the roll method shown in Fig. 5.3. Table 5.3 shows the main amorphous ceramics manufactured up to now using the roll method. From this table, three trends in the study of amorphous materials obtained by the roll method can be seen:

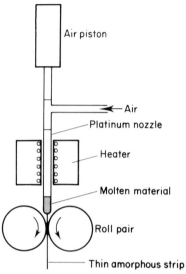

Fig. 5.3 Apparatus for making thin supercooled amorphous strips by the roll method

(i) the search for new glass formation systems, (ii) the development of functional materials such as amorphous ceramics that are strongly dielectric or show high ionic conductivity or electrochromism, and (iii) the study of crystallization processes.

What physical properties are brought about by the change to an amorphous structure? When a ferromagnetic material with a perovskite structure, which is highly symmetrical (little anisotropy) is made amorphous, the appearance of ferroelectric characteristics is theoretically possible. In reality, samples of amorphous $LiNbO_3$ have a very high dielectric constant at room temperature, and the unique characterisitic that has been found is its exhibiting a peak when a certain temperature is reached within the non-crystalline structure, as the temperature is raised. Whether this type of phenomenon reflects the unique non-crystalline structure or something else, something like a change in ionic conductivity, has not been established and still generates interest.

From an analogy with glass ceramics, good translucent electro-optical materials have been obtained by the crystallization of amorphous ceramics, since there are no pores or glass phases among the precipitated crystallites. Also, through control of the crystallization conditons, materials made up of smaller particles than those produced in the normal sintering reaction may be obtained.

Table 5.3 Amorphous materials obtained by ultra-rapid cooling

Composition	Characteristics
V_2O_5, TeO_2, MoO_3, WO_3	Uniform constituent glass formation [a]
$BaO-TiO_2$, $CaO-Al_2O_3$, K_2O-TiO_2 systems, etc.	2-constituent glass formation [a]
R_2O-MoO_3, R_2O-WO_3 systems (R = alkali)	2-constituent glass formation [a]
$BaO-Nb_2O_5$, $CeO_2-Nb_2O_5$, $La_2O_3-TiO_2$, $CeO_2-Al_2O_3$ systems, etc.	2-constituent glass formation
$LiNbO_3$, $LiTaO_3$	Ferroelectric character, high ionic conductivity
$LiNbO_3$, $PbTiO_3$	Highly dielectric, crystallization process
$Li_2O-Na_2O-K_2O-Ta_2O_5-Nb_2O_5$ systems	Glass formation, crystallization process
$Li_2O-M_2O_3$ systems (M = Al, Ga, Bi)	High ionic conductivity
$Li_2SO_4-La_2 (SO_4)_2$ systems	Glass formation
R_2O-WO_3, R_2O-MoO_3 systems (R = alkali)	High ionic conductivity, electrochromism
$Li_2O-BaO-Nb_2O_5$ systems	High ionic conductivity, glass structure
$LiNbO_3-Li_3PO_4$ systems	Structural analysis
$AgI-Ag_2P_2O_7$ systems	High ionic conductivity
$PbO-R_2O$, $PbO-R'O$ systems (R = alkali; R' = alkali earths)	Highly dielectric

[a] Splatter cooling, others are manufactured by the roll method.

Not only ferroelectric substances, but also ferromagnetic substances and ion conductors, etc., have been obtained in amorphous ceramic form. There is much promise for applications in optomagnetic fields and for various kinds of sensors, since these materials have excellent high-frequency characteristics and their being amorphous allows for high translucency when they are made into thin plates. The materials with high ionic conductivity shown in Table 5.3, especially, are being considered for oxygen-sensors. Also, there are expectations for uses in catalytic fields.

5.4 CAN CERAMIC SUPERCONDUCTORS BE PRACTICABLE?

The search for a new high critical temperature is an aspiration toward the unknown for mankind and a concrete demand of the current energy problem and information revolution. So far this search has been carried out in the so-called land of metals, and a map proceeding toward the interior through simple metals, alloys, interstitial compounds (carbides, nitrides) and intermetallic compounds is being made. The highest peak found in the land of metals is the critical temperature T_c of Nb_3Ge, which is 23.2 K. Other metals being used in practical materials are Nb–Ti alloy, Nb_3Sn, V_3Si, etc. And the eyes of the explorers have turned toward the land of inorganic substances such as oxides and sulfides, and the land of organic substances. In these lands, the second and third lands, the highest peak is still $PbMo_6S_8$ at 15 K, but there are many interesting mountains, and quite recently the discovery of a superconductor mountain in the organic land has become a big topic, even though its T_c is low.

Among the oxides there is a great number of materials that show the conductivity of metals, and if one were to show the superconductors, Table 5.4 would result. $SrTiO_{3-\delta}$, which was recently revealed, is a kind of degenerate semiconductor which is obtained through the introduction of oxygen vacancies by reduction, and its discovery was based on the theoretical possibility of superconductivity. Its T_c is low at 0.55 K, but even in oxides, it is possible to obtain a rather high T_c recently. These materials include spinel-structured $Li_{1+x}Ti_{2-x}O_4$ and a perovskite system solid solution of $BaPb_{1-x}Bi_xO_3$. Of these $Li_{1+x}Ti_{2-x}O_4$ is difficult to obtain in a single phase, and the synthesized samples undergo changes in quality at room temperature. And the latter, $BaPb_{1-x}Bi_xO_3$, decomposes at high temperatures (1,050 °C). The characteristic that both hold in common is that T_c is at its highest when formation X is changing, just before its properties shift from metallic to semiconductive.

In $BaPb_{1-x}Bi_xO_3$, the electron mobility μ becomes small near the boundary, and since the non-localized electron state has a lower limit of about $0.1 \ cm^2 \ S^{-1}V^{-1}$, the possibility that electron–phonon interaction will increase can be considered. However, as is shown by Fig. 5.4, even near the maximum T_c, the carrier density of this system is around $4 \times 10^{21} \ cm^{-3}$, which is a value about 10 times smaller than that for a standard high T_c superconductor.

Therefore, the possibility of a mechanism other than the standard BCS mechanism can be considered.

Table 5.4 Oxide superconductors at a glance

Name of substance	$T_c{}^a$	Structure	Year of discovery
$SrTiO_{3-\delta}$	0.55 (K)	perovskite	1964
TiO	2.3	NaCl	1964
NbO	1.25	NaCl	1964
M_xWO_3	6.7	bronze	1964
$Ag_7O_8{}^+X^-(X^- = NO_3{}^-,$	1.04	clathrate	1966
$F^-, BF_4{}^-)$			
$Li_{1+x}Ti_{2-x}O_4$	13.7	spinel	1973
$BaPb_{1-x}Bi_xO_3$	13	perovskite	1975

^{a} Highest for the substance or system.

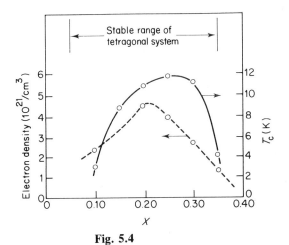

Fig. 5.4

Since it is difficult to manufacture these materials in single crystals, research must be carried out with polycrystalline samples in which there are uniformity problems at present. Therefore, there are many unresolved problems, but there are characteristics that are interesting from the practical point of view. Even if a $BaPb_{1-x}Bi_xO_3$ film is not sandwiched between insulating films, characteristics similar to those of a Josephson junction are produced at both ends, if a narrowly constricted, weakly bonded part is made. This is because the grain boundaries form an insulating barrier and can be thought of as playing the role of the insulating film. Therefore, it can be said that there are possibilities for developing devices in a form different from that used up to now with the unique characteristics of $BaPb_{1-x}Bi_xO_3$ films.

5.5 WHAT KINDS OF CERAMICS ARE THERE WITH GOOD THERMAL CONDUCTIVITY? WHAT ARE THEIR APPLICATIONS?

With the integration of electronic circuits and the creation of high-performance machinery continuing rapidly, the heat radiation technology for the accompanying heat generation has become an essential point in the functional upkeep of these apparatuses, and the development of ceramic insulators with excellent heat radiation characteristics is a big requirement. Since thermal conductivity is essentially determined by the elements and the bonding structure, the substances with possibilities as ceramics with high thermal conductivity are limited to the ones shown in Table 5.5. The ceramics that have been used as industrial materials include alumina (Al_2O_3), beryllia (BeO), boron nitride (BN), etc. However, BeO is a poisonous substance, BN has low strength, and the thermal conductivity of Al_2O_3 is limited, so there is no conclusive evidence as to which to use. Therefore there has been a rapidly growing tendency to develop new materials that have potential for high thermal conductivity.

Table 5.5 Comparison of characteristics for various materials

Material	Thermal conductivity (W m^{-1} K^{-1})	Thermal expansion coefficient ($\times 10^{-3}$degC^{-1})	Resistivity (Ω cm)	Bending strength (MN m^{-2})
SiC–BeO system ceramics	270	3.7	>4×10^{13}	450
Si$_3$N$_4$ ceramics	13–55	2.8–3.2	>10^{14}	390–1,300
BN ceramics (95%)	57	−0.7	>10^{14}	24
BeO ceramics (99%)	240	68	>10^{14}	200
Al$_2$O$_3$(99.5%)	29	72	>10^{14}	290
Al	230	257	2.7×10^{-6}	—
Cu	390	175	1.7×10^{-6}	—
Single-crystal Si	125	3.5–4.0	—	—

Recently, a silicon carbide (SiC) ceramics that has thermal conduction and electrical insulation properties similar to BeO has been developed. In general, thermal conductivity has been poor in substances which are highly insulating, and conversely, resistance has been low in substances with high thermal conductivity up until now, and SiC ceramics was not an exception. However, recently, SiC ceramics with a BeO additive has been developed, and compared with the material up to now, it has four times the thermal conductivity and a resistivity which is larger by 10 decimal places. As can be understood from Table 5.5, it combines a thermal conductivity of 270 W/mK which surpasses Al and a resistivity of 4 × 10^{13} Ω cm, which comes near to that of Al_2O_3 ceramics.

Why, then, can this strong insulating property and high thermal conductivity coexist in SiC–BeO system ceramics? The key is in the BeO additive. The SiC crystallites are n-type semiconductors. When a mixture of SiC and BeO powders is sintered, the resistance within the SiC crystallites is low after the sintering, but in the grain boundaries, the resistance is high, the exact same functional mechanism as is used in the well-known ZnO varistor. Up to now, SiC ceramics has had large amounts of impurities which formed a low-resistance impurity layer in the grain boundaries. On the other hand, even though BeO hardly enters into the crystallites at all and is precipitated into the grain boundaries, the layer is extremely thin and there is almost nothing in the grain boundaries. BeO itself has a large resistivity and high thermal conductivity, and these make it work even better.

In insulators like SiC ceramics, lattice vibrations (phonon) conduct heat, but there must be no impurities or pores that increase the thermal resistance through phonon-scattering, in order to facilitate a low thermal resistance operation.

One can see that these new ceramics will exert an influence on the substrates for LSI (large-scale integrated circuit) chips, where the heat radiation problem began, and on high density hybrid IC substrates, etc. In Fig. 5.5, a hybrid circuit that uses the alumina substrate used up to now and one that uses a new ceramic substrate are shown. In order to show the main point only in this figure, the resistors and condensers have been omitted. It should be evident from Fig. 5.5(b) that since it is extremely simple, the new ceramics construction is not only advantageous in terms of the effects of the number of parts, assembly processes, and amount of assembly time and cost, but also in terms of lightweight, miniaturization, and reliability.

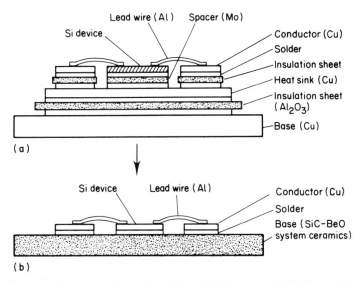

Fig. 5.5 Structural cross-sections of modules. (a) Structure of Al_2O_3 modules (up to present). (b) Structure of SiC–BeO ceramics module

5.6 WHAT KINDS OF GLASS ARE USED FOR MEDICAL GLASS (BIOGLASS)?

Metals have been used in the treatment of fractured bones for some time, and dentists have used ceramics in the same ways they use precious metals for fillings after the treatment of cavities. Also, proper use of the various types allows for matching the colors of the ceramic teeth with those of the natural ones. In terms of the function of inorganic substances in contact with the human body, it can be said that there is no reaction; that is, they can be considered to exist in chemical stability. Compared with the stable precious metals and ceramics, glass is a little unstable, and it tends to dissolve with prolonged immersion in water, acid, or alkali and to form an intermediate layer. It is commonly known that in commercial glass, the constituents such as alkali ions, which are easily dissolved in water, flow out, and a silica gel layer is left on the surface of the glass. Bioglass is a product of the research on the inorganic substances that can form structures which adapt to the living body by means of a positive matching of this instability with the growth processes of tissues. Some typical examples of the bioglasses that have been studied are shown in Table 5.6. They characteristically contain phosphoric acid and calcium, and their modified oxide meshes with large amounts of Na and Ca explain why they are more unstable than commercial glass. When glass with these compositions is inserted into the living body, there is a reaction with the body fluids, and the Na^+ ions in the surface part of the glass are flowed out.

Table 5.6 Examples of bioglass compositions

Constituent	45S5	45S10	45S5F	45B5S5
SiO_2	45	45	45	45
Na_2O	24.5	27.5	23	24.5
CaO	24.5	24.5	12	24.5
CaF_2	—	—	16	—
B_2O_3	—	—	—	5.0
P_2O_5	6.0	3.0	6.0	6.0

Then a silica gel which centers around an $-\overset{\displaystyle |}{\underset{\displaystyle |}{Si}}-OH$ structure forms on the surface of the glass. Next, collagen fibers begin to grow from the living tissue and, as is shown in Fig. 5.6, a combined tissue is formed. In this way a bond layer is formed between the living tissue and the surface of the glass, and it becomes a strong inorganic cement between the bone and the artificial tissue that has been implanted into it. Also, metals and alumina ceramics with a bioglass glaze which is applied through a coating process have been studied for implantation in the body as bone replacements. As a result of experiments with animals, it has been confirmed that 3 to 4 weeks after implantation in the body, the growth cells of the bone begin making fibrous collagen in the gel layer, and there is a new synthesis of $Ca_5(PO_4)_3 \cdot OH$ hydro-oxy apatite, for which the glass supplies the Ca and P. Torsion tests have confirmed a mechanical strength greater than that of the bone itself after half a year.

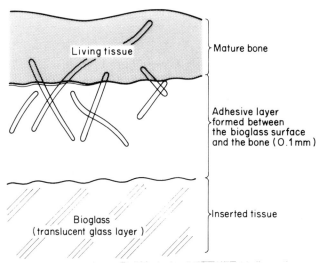

Fig. 5.6 Interface between bioglass and living tissue

It has also been confirmed that when the glass is filled with a large number of tiny pores (porous product of Bicheroux acid treatment), this high-porosity substance forms a complete adhesive layer in 4 weeks; after that, the filled layer is gradually absorbed by the body, and hydro-oxy apatite begins to grow

Bioglass is used independently for spinal fusions and it has the effect of protecting the spinal cord. It is also used as a tooth implant material in which stainless steel, Co–Cr–Mo alloy, Ti, Ti alloys, or Al_2O_3 is coated with it. The material mentioned above, along with bioglass–metalfiber, is being investigated as a means of repairing complex fractures and is used in screws and wires. It also seems that there will be surgical uses for fine ceramics.

5.7 EXPLAIN ULTRA-FINE CERAMIC POWDER

The size of devices in the fine fabrication processes typical of IC technology is of the order of micrometers. The manufacturing and advance of usage technology of these kinds of miniature devices is limited at present to materials that center on several semiconductors, Si, Ge and GaP. In order for the miniaturization of electronic devices to attain high performance and reliability system-wide, fine fabrication of other surrounding devices is necessary, and in the near future technological developments are necessary.

The typical fine fabrication method being put into practice is the particle integration method. This method handles particles or microparticles on the level of the smallest chemical constituent, the atom or molecule, and integrates them into the material. With this method it is possible to manufacture miniature ceramic devices in sizes about the same as the particles. The device's attainable dimensions depend on the size of the particles: the smaller the particles are, the more miniature the solid device becomes. Therefore, manufacturing particles with sizes that suit the desired ceramics is a prerequisite.

One of the techniques that it is thought will become one of the most effective for fine particle powders is manufacture from metal alkoxides. As can be seen in Fig. 5.7, the use of alkoxides in this method allows one to consider it as one of the ceramic powder synthesis methods that employ solution technology. The alkoxide process is one that can be planned for good function from the raw materials to the products by means of particle integration throughout the whole process. It is a process for the next century that points out the direction of modern science and technology, because it is (i) a low energy process, (ii) a process in which the reactions take place under mild conditions like those of the body, and (iii) a clean process in which the synthesis of the substances conforms to the materials made. Fig. 5.7(b) is a flowchart showing the manufacture of $BaTiO_3$ powder from the metal alkoxide.

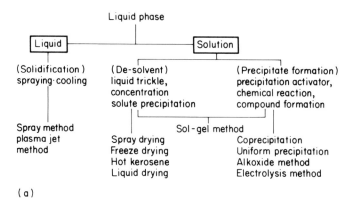

(a)

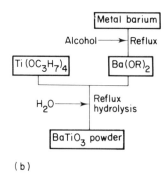

(b)

Fig. 5.7 (a) Powder regulation from liquids. (b) Flowchart for production of $BaTiO_3$ powder by the alkoxide process

The alkoxide process can be considered to be an important micrometer powder fabrication technology for the near future. However, no matter how excellent the process is, the impact on industry will not be important if the applications are limited to a few specific ceramics.

Based on the progress in alkoxide chemistry during the past 30 years, the synthesis of the alkoxides of most elements in the periodic table is possible. That is to say that all of the metallic elements that go into forming ceramics can be obtained from alkoxides.

The 'integrated ultra-fine particle gas-sensor' is an example of the applications of ultra-fine particles. With standard gas-sensors, if the active element is not heated to several hundred °C, the necessary gas-sensitivity and response speed cannot be obtained, but the monolithic integrated circuit chips which make up the microprocessor elements necessary for integration cannot generally be operated at temperatures above 130–150 °C. Since fine particle gas-sensitive film works at low temperatures of 130–150 °C, a single-chip, monolithic-type, integrated ultra-fine particle gas-sensor which includes a microprocessor element is possible using this film and 'integration' (Fig. 5.8).

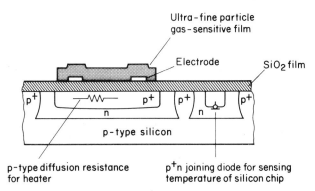

Fig. 5.8 Cross-section of a integrated ultra-fine particle gas-sensor prototype

5.8 EXPLAIN THE TECHNIQUES FOR LAMINATED AND MULTILAYER CERAMICS

Along with ICs and LSIs, the miniaturization and integration of parts have progressed rapidly, and besides laminated condensers, one comes across laminated batteries, laminated piezoelectric products, laminated chip coils, laminated substrates, laminated sensors, etc., all terms in which the word 'laminated' appears. This is because all of them are basically a piling of thin layers, the whole of which is intended to produce miniaturization and a new range of characteristics. If one looks at these laminated parts in terms of single-layer functions, they divide roughly in two. In one group, the function of each single layer is in principle the same as each other layer, and a quantitative increase in that function is produced in proportion to the number of layers. The other group is one in which an accumulation of single layers with different functions produces yet another function; and since these laminated devices can have multiple functions, this type is appropriate for laminated substrates and laminated sensors. Laminated condensers, laminated varistors, and laminated piezoelectric products are examples of the type mentioned first. Of these, laminated varistors and laminated substrates will be discussed. Fig 5.9 shows a structural schematic of a laminated condenser.

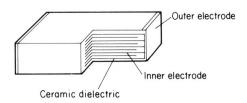

Fig. 5.9 Structural schematic of a laminated condenser

(1) Laminated varistor

In order to lower the threshold voltage $V_{1\,mA}$, miniaturize the form and improve the anti-surge capacity, this varistor is made with the same structure as a laminated condenser by means of a green sheet method. Fig. 5.10 shows the relationship between the thickness of the green sheet and $V_{1\,mA}$. As is evident from this figure, they are proportional to each other and $V_{1\,mA}$ can be controlled through the thickness of the green sheet. With a 40 μm thick green sheet, a low $V_{1\,mA}$ of 4.2 V can be obtained. Fig. 5.11 shows an example of the current vs. voltage characteristics for a laminated varistor. In this example, the non-linearity coefficient $\alpha = 30$ to 38 is a large improvement over an equivalent single-piece varistor's $\alpha < 20$.

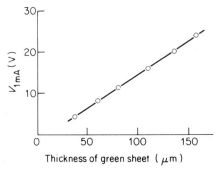

Fig. 5.10 Relationship between green sheet thickness and threshold voltage in a laminated varistor

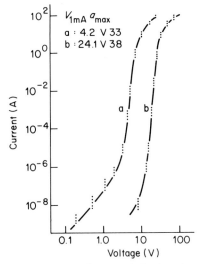

Fig. 5.11 Voltage vs. current characteristics in a laminated varistor

(2) Laminated substrates

ICs and LSIs have been developed, and now there is a rush into the VLSI age. In association with this, the operational delay time in logic circuits has been greatly reduced, so the signal propagation delay time in distribution wires and the delay time associated with loads have become bigger problems. That is, the delay time associated with mountings has come to be a controlling factor of system-wide speed. Because of this, high level integrated mounting technology is necessary.

In the logic circuits of the processor unit of the IBM 3081 large-scale computer, 100 or 118 flip chip LSIs have been loaded onto the substrate with input/output pins coming out the back. A 90-mm square ceramic substrate with a maximum of 33 layers is used. The fabrication method for these laminated substrates involves making holes for through-hole use in the green sheet with a high-speed punch, printing a molybdenum paste, laminating, and sintering. In order to connect the through holes accurately, there is a necessity for high-level control of the shrinkage rate of the green sheet. Because of this, control of the raw materials, intermediate processes, and final processing is extremely important.

Index